DICTIONARY OF ELECTROCHEMISTRY

Second Edition

DICTIONARY OF ELECTROCHEMISTRY

Second Edition

D. B. Hibbert

and

A. M. James

MACMILLAN PRESS
LONDON

Macmillan Reference Books

© D. B. Hibbert & A. M. James, 1984

All rights reserved. No reproduction, copy or transmission of this publication may be made without written permission. No paragraph of this publication may be reproduced, copied or transmitted save with written permission or in accordance with the provisions of the Copyright Act 1956 (as amended). Any person who does any unauthorized act in relation to this publication may be liable to criminal prosecution and civil claims for damages.

First edition published 1976

Second edition first published 1984 by
THE MACMILLAN PRESS LTD
London and Basingstoke

Associated companies in Auckland, Delhi, Dublin, Gaborone, Hamburg, Harare, Hong Kong, Johannesburg, Kuala Lumpur, Lagos, Manzini, Melbourne, Mexico City, Nairobi, New York, Singapore, Tokyo.

British Library Cataloguing in Publication Data

Hibbert, D.
  Dictionary of electrochemistry.—2nd ed.
  1. Electrochemistry—Dictionaries
  I. Title II. James, A. M.
  541.37′0321   QD552.5

ISBN 0-333-34983-0

Photoset by Paston Press, Norwich
Printed in Great Britain at The Pitman Press, Bath

PREFACE

This is a dictionary of electrochemical terms and concepts, and is intended as a reference volume for physical and biological scientists in education, research and industry.

Electrochemistry, although formally a branch of physical chemistry, now has a wider significance and application in such areas as the biological sciences, physics, metallurgy and engineering. The first edition of this dictionary has been extensively revised and largely rewritten to include recent developments in such fields as fuel cells, corrosion, energy conversion, electrode kinetics, ion-selective electrodes and bioelectrochemistry. The aims are still the same, namely to provide the reader with a brief and lucid definition of an unfamiliar term, its significance in various branches of science and, if it is a measurable quantity, a concise account of its measurement. Electrochemical and thermodynamic formulae and equations are quoted without detailed proof, but their applications and limitations are discussed. The 360 entries are arranged in alphabetical order, many being illustrated with line diagrams and cross-referenced for easy access. Bibliographical references to standard texts and review articles give access to background information and further reading. SI units are used throughout and the necessary conversion factors to other units are provided. A selected tabulation of electrochemical data is included.

It is hoped that this dictionary, with its extended coverage, will provide a ready reference for biochemists, biologists, chemists, geologists, microbiologists, pharmacists, physicists and applied scientists. Although it is not intended as a textbook, it will be useful to first-year university-level students by providing them with basic information on the important concepts of electrochemistry. Second- and third-year students will welcome it as a revision source of electrochemical principles.

We are indebted to Dr. Ann Ralph for her practical help, advice and useful discussion during the production and editing of the manuscript. Our sincere thanks are due to our spouses Mary and Marion for their understanding and patience during the writing of this dictionary.

AMJ
DBH

HOW TO USE THIS DICTIONARY

Entries are arranged in alphabetical sequence with full cross-referencing where there is more than one acceptable or recognized name.

Symbols and abbreviations are used in the text without definition; reference should be made to the list of Principal Symbols (*see* p. vii).

In the text, words in **bold** type indicate a reference to another entry which would be of help; in some instances the reference to another entry is in parentheses. Thus in the sentence 'Diffusion near the electrode surface gives rise to concentration overpotential (see **Overpotential**), which in turn leads to the **limiting current density**', the reader is referred to entries 'overpotential' and 'limiting current density' for further information.

More detailed treatments of some of the entries can be found in standard reference books; where appropriate, relevant references are indicated at the end of an entry. All the books, together with a compilation of journals on electrochemistry and related subjects, are listed in the Bibliography.

Formulae used as subscripts are written in linear form, thus:

a_{Cl-}, a_{H3O+} and a_{Ca2+} represent the activity of the chloride ion, the solvated proton and the calcium ion, respectively;

$a_{Cl-,1}$ and $a_{Cl-,2}$ represent the activity of the chloride ion in two different solutions or states 1 and 2, respectively;

$a_{\pm,1}$ and $a_{\pm,2}$ represent the mean ionic activity of a specified electrolyte in two different solutions designated 1 and 2, respectively.

PRINCIPAL SYMBOLS

A	area; Debye–Hückel constant; absorbance	G	Gibbs free energy function, conductance
A	ampere	$\bar{G}$	electrochemical free energy
A_r	relative atomic mass		
a	mean ionic diameter	g	gram
a_A, a_B, a_i	activity of A, B or ith component	H	enthalpy
		Hz	hertz
$a_+, a_-, a_\pm$	activity of cation, anion, mean ionic activity	h	Planck's constant
		I	ionic strength; electric current
B	Debye–Hückel constant	i	current density
		i_0	exchange current density
C	capacitance		
$\tilde{C}$	differential capacitance	i_L	limiting current density
C	coulomb	J	joule
c_A	concentration of A/mol dm^{-3}	J	cell constant
		K	equilibrium constant
D, D_+, D_-	diffusion coefficient of cation, anion	K_a, K_b	acid and base ionization constants
E	electromotive force	K_w	ionization constant of water
$E_{X+,X}$	electrode potential		
$E_{ox,red}$	oxidation–reduction potential	K	kelvin
		k	Boltzmann constant
E'	formal electrode potential	k_f, k_b	rate constants for forward and back reactions
$E^{\ominus\prime}$	standard electrode potential at pH 7.0		
		L	length
$E_{1/2}$	half-wave potential	l	length
e	electron, electronic charge	M_r	relative molecular mass
		m_A	molality of A
F	Faraday constant		
F	farad		

Principal symbols

$m_+, m_-, m_\pm$	molality of cation, anion, mean ionic molality	α	degree of dissociation, association; transfer coefficient
m	metre	β	symmetry coefficient
N	newton	$\gamma_i, \gamma_+, \gamma_-, \gamma_\pm$	activity coefficient of ion, cation, anion, mean ionic activity coefficient
N_A	Avogadro constant		
n	number of electrons transferred		
n_A, n_B	number of molecules of A, B in the system	Δ	increase in thermodynamic function, e.g. ΔG
P	total pressure of system	δ	thickness of diffusion layer
p_i	partial pressure of i in system	ε	efficiency
Q	electric charge	$\varepsilon, \varepsilon_0, \varepsilon_r$	permittivity, permittivity of free space, relative permittivity (dielectric constant)
R	resistance; gas constant		
S	entropy; coefficient in Onsager's equation		
S	siemens	ζ	zeta, electrokinetic potential
s	second		
T	temperature/K	η	viscosity coefficient, overpotential, overvoltage
t	time		
t_+, t_-	transport number of cation, anion	θ	fraction of surface covered
u, u_+, u_-	mobility of ion, cation, anion	κ	conductivity, reciprocal thickness of double layer, transmission coefficient
V	potential; volume		
V	volt		
v, v_+, v_-	velocity of ion, cation, anion	Λ	molar conductivity
		$\Lambda_i, \Lambda_+, \Lambda_-$	molar conductivity of ion, cation, anion
W	watt	μ_A, μ_B	chemical potential of A, B
w_A, w_B	weight of A, B		
X	applied field strength	$\bar{\mu}_A, \bar{\mu}_B$	electrochemical potential of A, B
x_A, x_B	mole fraction of A, B	ν, ν_+, ν_-	number of ions, cations, anions formed from 1 mol of electrolyte
Z	impedance		
z_i, z_+, z_-	charge number of ion, cation, anion		
		ν	stoichiometric number, frequency

Principal symbols

ξ	reaction coordinate	P, V, T, S, η	indicating constant pressure, volume, temperature, entropy, overpotential
π	ratio of circumference to diameter of circle		
ρ	density, resistivity		
σ	surface charge density	$+, -$	referring to positive, negative ion
τ	transition time		
ϕ	inner or Galvani potential	1, 2	referring to different systems or states of system
χ	surface potential		
ψ	outer or Volta potential		
Ω	ohm	**Other abbreviations**	
ω	angular velocity	g, l, s	referring to gaseous, liquid or solid state
Superscripts		aq	aqueous
$\ominus$	indicating standard value of a property	bp	boiling point
		mp	melting point
o	indicating value of a property at infinite dilution	vp	vapour pressure
		pH	$-\log a_{H^+}$
		pK	$-\log K$
$\ddagger$	indicating value of a property in the transition state	**Mathematical symbols**	
		$\approx$	approximately equal to
N	oxidation state	$\simeq$	asymptotically equal to
		$\sum_i$	sum of i terms
Subscripts			
A, B ...	referring to species A, B ...	$\prod_i$	product of i terms
		$\ln x$	natural logarithm of x
e	value of a property at equilibrium	$\log x$	common logarithm of x (to base 10)
i	referring to typical ionic species i	$|x|$	absolute value of x

A

Accumulator. *See* **Electrochemical storage.**

Acids and bases. According to the Lewis–Brønsted definition, an acid is a substance that can donate a proton and a base is one that can accept a proton

$$\text{acid} \rightleftharpoons \text{proton} + \text{base}$$

An acid can only give up a proton if a base is present to accept it, hence two acids and two bases take part in any reaction involving proton transfer, that is

$$HA_1 + B_2 \rightleftharpoons HA_2 + B_1$$

Since free protons do not exist in solution, the functioning of an acid depends on the presence of a base to which protons can be transferred. The solvent thus plays an important role in determining acidic or basic properties of a substance.

A solvent SH capable of being both a proton donor and acceptor (i.e. amphiprotic) can undergo autoprotolysis

$$SH + SH \rightleftharpoons SH_2^+ + S^-$$

for which

$$K_s = a_{SH2+} a_{S-}$$

Since pure solvents have different molalities, a better comparison of autoprotolysis constants can be obtained by dividing K_s by the square of the molality of the pure solvent

$$K'_s = \frac{K_s}{m_{SH}^2} = K_s \times M^2$$

where M kg is the molar mass of SH. K_s values are determined from conductance or e.m.f. data (*see* **Ionic product of water**). There is no correlation between pK_s or pK'_s and the dielectric constant of the solvent (*see* Table 1).

In such solvents, the strongest acid is the solvent cation SH_2^+ and the strongest base the solvent anion S^-. In water, strong acids (bases) are converted completely to the conjugate base (acid) and the solvent cation H_3O^+ (anion OH^-) and are said to be levelled. In other solvents, intrinsic

2 Acids and bases

Table 1
Autoprotolysis constants at 298 K

Solvent	pK_s	pK'_s	Dielectric constant
Sulphuric acid	3.57	5.59	101
Ethanolamine	5.7	8.1	37.7
Water	14.00	17.49	80.36
Acetic acid	14.5	16.9	6.1
Methanol	16.6	19.6	33.7
Ethanol	18.8	21.5	24.5
Ammonia[a]	29.0	32.0	22.0
Acetonitrile	>32	>35	37

[a] At 240 K.

differences in acid (base) strength may not be levelled because of incomplete dissociation, and the acids (bases) can be differentiated and listed in order of strength. A strongly acidic solvent (e.g., sulphuric acid) levels all bases converting them completely to their conjugate acid and HSO_4^-. A basic solvent (e.g., ethanolamine) levels many acids, converting them to the solvent cation ($HOC_2H_5NH_3^+$) and will differentiate many acids.

The value of the autoprotolysis constant sets the limit of differentiation of the solvent. For water acids with pK values in the range 1–13 can be studied, for sulphuric acid the range is 0–3.6 whereas for acetonitrile the range is 0–32; the weakest acid so far investigated is phenol ($pK = 26.6$ in acetonitrile). The autoprotolysis constant is important in determining the sharpness of an acid/base titration; solvents with small values of K_s give the sharpest endpoints.

In hydrogen-bonded solvents (protic) of high dielectric constant, the reaction of an acid is represented by

$$HA + SH \rightleftharpoons SH_2^+ + A^-$$

and a base by

$$B + SH \rightleftharpoons BH^+ + S^-$$

in which all the species are solvated, principally by hydrogen bonding with the solvent. Ion pairing is negligible, except in methanol. Reactions of this type proceed further in the forward direction in a solvent such as water than in methanol or ethanol because less work is required for the dissociation process. The magnitude of this effect can be large: for example, K_a for acetic acid in water is 1.75×10^{-5} mol dm^{-3} and in methanol is 1.95×10^{-10} mol dm^{-3}. The medium effect measured by

$$\Delta pK = pK_{a,\text{solvent}} - pK_{a,\text{water}}$$

is approximately constant for a given class of acids in one of these solvents.

Non-hydrogen-bonded (dipolar aprotic) solvents of high dielectric constant (e.g., nitrobenzene, N,N-dimethylformamide, acetonitrile, nitromethane) show little ability to donate protons to bases. All such solvents are sufficiently basic to combine with and stabilize protons from acids. Molecules of these solvents are more polar than those of protic solvents and thus there will be stronger ion–dipole and molecule–dipole solute–solvent interactions.

Hydrogen-bonded solvents of low dielectric constant (<30) (e.g., ethanol, higher alcohols, ethylene diamine, acetic acid) are capable of donating a hydrogen atom to hydrogen bonds. Anions are more solvated and stabilized by such hydrogen-bond donors than by dipolar aprotic solvents. Ion pairing is an important complication in acid/base behaviour in such solvents. In ethanol, the force between two ions (a specified distance apart) is three times that in water and twice that in dimethylsulphoxide; in acetic acid, the force is four times that in ethanol. The strength of an acid in these solvents depends on the base with which it reacts, whether this is the solvent or some basic solute. There is no unique scale of acid strength in solvents of low dielectric constant.

In non-hydrogen-bonded solvents of low dielectric constant (e.g., acetone, benzene) acid/base reactions are complicated by extensive ion pairing and by the formation of other ionic and molecular aggregates. Solvents of very low dielectric constant (benzene, $\varepsilon_r = 2.29$) are not sufficiently basic to remove protons from acids and so acid/base behaviour is confined to association reactions between acidic and basic solutes of the type

$$B + HA \rightleftharpoons BHA$$

Association usually involves ionization to produce a hydrogen-bonded ion pair BH$^+$ A$^-$, but dissociation to the free ions can be neglected. The quotient

$$K_{\text{BHA}} = \frac{a_{\text{BHA}}}{a_\text{B} a_{\text{HA}}} \approx \frac{c_{\text{BHA}}}{c_\text{B} c_{\text{HA}}}$$

is referred to as an association constant rather than an ionization constant. Association constants may be obtained from spectroscopic studies; the strengths of bases can be compared by means of their association with an indicator such as 2:4-dinitrophenol. See Covington and Dickinson (1973), chapter 3.

Ac impedance methods. Techniques in which the resistance and capacitance of an electrode, or cell, are measured often in the presence of a dc bias. These methods are capable of high precision and may be used in the evaluation of

4 Ac impedance methods

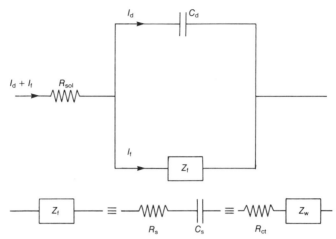

Figure 1. An electrical circuit which is equivalent to an electrochemical cell. Z_W is the Warburg impedance of the charge transfer reaction

heterogeneous charge transfer parameters and for studies of the structure of the double layer.

An electrochemical cell may be described in terms of an equivalent electrical circuit (*see* Figure 1) which contains elements due to the double layer structure (the capacitance of the double layer C_d through which the current I_d flows), the faradaic process (impedance Z_f through which I_f flows) and the solution resistance (R_{sol} through which $I_d + I_f$ flows). The faradaic impedance is equivalent to the sum of a series resistance R_s and a capacitance (the **pseudocapacitance** C_s) or Z_f may be written as the sum of a charge transfer resistance R_{ct} and an impedance Z_W, the Warburg impedance. The faradaic impedance is measured as a function of the ac frequency, and the variation of R_s and C_s (or R_{ct} and Z_W) discussed in terms of the electrochemical processes occurring in the cell. A complex plane plot of the real and imaginary parts of the impedance is shown in Figure 2. While the system is under kinetic control, the plot is semicircular cutting the imaginary axis at R_{sol} and $R_{ct} + R_{sol}$. As mass transfer becomes important at lower frequencies a linear plot is found. Values of the standard heterogeneous rate constant of the electrochemical reaction lying between 1 cm s^{-1} and 2×10^{-5} cm s^{-1} may be determined from measurement of R_{ct} ($R_{ct} = RT/nFI_0$).

The imposition of a ramped dc bias on the ac signal leads to ac voltammetry or, if a dropping mercury electrode is used, to ac polarography (*see* **Polarography**).

Ac impedance measurements may be made quickly with the aid of a wide-band excitation signal. A fast Fourier transform of the current and voltage output leads to a distribution of the harmonics which comprise the

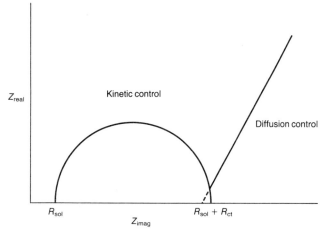

Figure 2. Complex plane analysis of the faradaic impedance of an electrode reaction showing regions of kinetic and diffusion control

signal. Thus a complete frequency scan may be obtained from a single measurement. See Bard and Faulkner (1980).

Activation overpotential. *See* **Overpotential.**

Activity (a). A function introduced by G. N. Lewis to aid the treatment of real systems. Changes in the **chemical potential** (μ) can be correlated with experimentally measured quantities through a relationship formally equivalent to that for an ideal system

$$\mu_i - \mu_i^\ominus = RT \ln a_i = RT \ln m_i \gamma_i$$

where γ_i is the activity coefficient.

Since no method has been devised to determine individual ionic activities (a_+ and a_-), the mean ionic activity ($a_\pm$) for dissolved electrolytes is defined as

$$a_\pm = a_2^{1/\nu} = (a_+^{\nu+} a_-^{\nu-})^{1/\nu}$$

where a_2 is the activity of the undissociated electrolyte.

Values of the mean ionic activity for electrolytes can be obtained from measurements of the e.m.f. of a **concentration cell** or **reversible galvanic cell**.

Activity coefficient (γ). The ratio of the **activity** to the concentration of the component i in the given state (i.e. $\gamma_i = a_i/m_i$). The activity coefficient, which approaches unity as the concentration approaches zero, is a measure of the departure of the system from ideal behaviour.

6 Adiponitrile synthesis

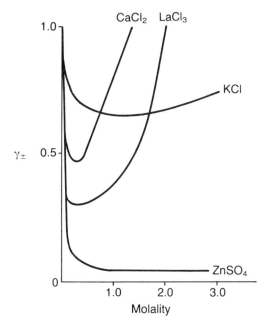

Figure 3. Activity coefficients of electrolytes of different valence type

For electrolytes γ_+ and γ_- cannot be measured separately, hence the mean ionic activity coefficient $\gamma_\pm$ is defined as

$$\gamma_\pm = (\gamma_+^{\nu+} \gamma_-^{\nu-})^{1/\nu}$$

$\gamma_\pm$ decreases rapidly with increasing molality and at higher concentrations usually passes through a minimum (*see* Figure 3). The steepness of the initial decrease varies with the valence type of the electrolyte, the greater the product of the charge carried by the ions the greater the deviation from ideal behaviour (*see* Figure 4). In dilute solution, activity coefficients can be calculated using the **Debye–Hückel activity equation**.

Mean ionic activity coefficients may be determined from a study of the e.m.f. of suitable cells with the concentration of the electrolyte solution (*see* **Electrode potential**).

Adiponitrile synthesis. *See* **Organic synthesis.**

Adsorption at an electrode. Both ions and neutral molecules may adsorb at an electrode in a process that is usually potential-dependent. Contact adsorption of ions (*see* **Electrical double layer**) can have a profound effect on the electrical

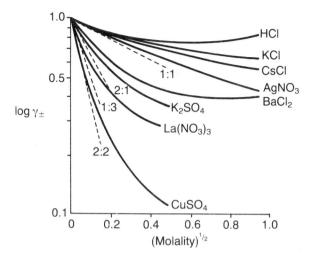

Figure 4. Mean ionic coefficients as a function of molality. (Dotted lines, behaviour expected from Debye–Hückel limiting law)

characteristics of an electrode. A layer of adsorbed ions at the inner Helmholtz plane constitutes, with the electrode surface and ions at the outer Helmholtz plane, a triple layer. In order to explain the shapes of curves of the double layer capacity against electrode charge, Bockris proposed an isotherm for the adsorption of ions

$$\ln \frac{\theta}{1-\theta} = A + \ln a_\pm + Bq + C\theta^{3/2}$$

where A, B and C are constants, and q is the electrode charge density. It may be concluded that simple isotherms which work well in the gas phase, such as those of Langmuir and Temkin, cannot be used to describe ionic adsorption at electrodes. Neutral species may also adsorb at an electrode. Indeed, electrodes in aqueous electrolyte are covered with water molecules which must be displaced before other species (e.g., organic molecules) may adsorb. A major contribution to the variation of adsorption with electrode potential comes from the effect of the electrode charge on the dipolar water layer. Consideration of the forces between the electrode and water molecules suggests that the water will be most weakly bound near the **potential of zero charge** and therefore the greatest adsorption of organic molecules will occur.

The formation of an adsorbed intermediate during an electrochemical reaction leads to an apparent increase in the double-layer capacitance caused by the charge required to form the adsorbed layer (*see* **Pseudocapacitance**).

8 Air electrodes

The extent of adsorption, which may be measured by various **transient methods** (*see also* **Linear sweep voltammetry; Stripping voltammetry**), can be used as a mechanistic indicator (*see* **Electrode reaction mechanisms; Hydrogen electrode reactions**). See Bockris and Reddy (1973).

Air electrodes. *See* **Metal–air cells.**

Alcohol–air cell. *See* **Fuel cell.**

Alcohol meter. A **fuel cell** using alcohol vapour as fuel constructed such that the current passed is proportional to the concentration of alcohol. The electrodes are constructed of platinum dispersed on graphite and mixed with polytetrafluoroethylene (PTFE) in the form of a **Teflon-bonded electrode**. The cathode is open to air and the electrolyte is dilute acid.

When used as a breathalyser 1 cm^3 of breath is allowed to contact the anode. No potential is applied, thus the current output is essentially at the **open-circuit voltage**. The total amount of alcohol in the sample is proportional to the area under the current–time curve (*see* Figure 5). It has been found that the peak current is also proportional to the concentration of alcohol, and this value is the one taken in a breath-testing apparatus. Carbon monoxide,

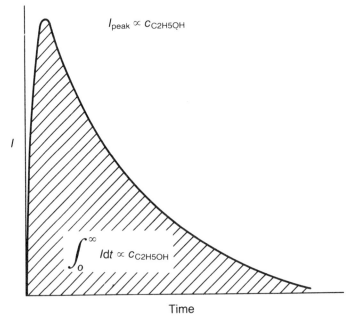

Figure 5. Current–time response of a breathalyser-type alcohol meter

hydrogen and acetone do not interfere with the determination. The cell reactions are

cathode	$3O_2 + 12H^+ + 2e \rightarrow 6H_2O$
anode	$C_2H_5OH + 3H_2O \rightarrow 2CO_2 + 12H^+ + 12e$
overall	$C_2H_5OH + 3O_2 \rightarrow 2CO_2 + 3H_2O$

See Pletcher (1982).

Alkaline cell. Primary cell using sodium or potassium hydroxide as the electrolyte. Alkaline cells based on a manganese dioxide cathode and zinc anode have a lower internal resistance and consequently better performance than the older **Leclanché cell**. In particular at high current drain, a higher potential is maintained for longer. Also, the shelf life of this cell is longer and a higher **energy density** may be achieved. Figure 6 shows a comparison between an alkaline cell and a Leclanché cell under constant current drain of 500 mA.

The stoichiometry of the overall electrode reactions is

cathode	$2MnO_2 + 2H_2O + 2e \rightleftharpoons Mn_2O_3 \cdot H_2O$	$E^\ominus = +0.118$
anode	$Zn + 2OH^- \rightleftharpoons Zn(OH)_2 + 2e$	$E^\ominus = -1.245$
overall	$Zn + 2MnO_2 + 2H_2O \rightleftharpoons Mn_2O_3 \cdot H_2O + Zn(OH)_2$	$E^\ominus = 1.433$

In practice, the potential given by the cell is about 1.9 V.

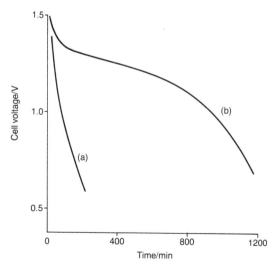

Figure 6. Voltage–time curves for (a) a Leclanché cell and (b) an alkaline manganese dioxide–zinc cell

10 Alkaline manganese battery

A **button cell** may be constructed and, with some changes in the structure of the battery, alkaline manganese dioxide–zinc cells may be recharged.

Alternative anode materials are cadmium, indium, aluminium or sodium. Alternative cathode oxides are those of copper, iron, lead, mercury, nickel or silver.

Alkaline manganese battery. *See* **Alkaline cell.**

Alloy electrodeposition. The commonest alloy to be electroplated is brass, which may be obtained in a wide range of copper/zinc ratios from cyanide baths. Surface coatings consisting of alloys of copper and cadmium, or containing tin, zinc and nickel, have also been produced.

The general considerations are the same as for single metals (*see* **Electroplating**), but close control of the bath is necessary if variations in the alloy composition are to be avoided. The best conditions, and the composition of the bath, are determined empirically.

During deposition of two metals (denoted 1 and 2), assuming the surface to be equipotential ($\Delta\phi$) with respect to the bulk solution, the following equalities must hold

$$\Delta\phi = \Delta\phi_{1,e} + \eta_1 = \Delta\phi_{2,e} + \eta_2$$

where $\Delta\phi_e$ is the reversible potential. Thermodynamic factors that influence the rate are contained in the $\Delta\phi_e$ terms. These depend not only upon the activities of the ions in solution, but also on the composition of the alloy. The kinetic aspects of the deposition rate are given by the **overpotential** terms η and are related to the deposition currents via the **Butler–Volmer equation**, in which the **exchange current density** of each metal deposition largely determines the facility of the process.

Aluminium electrometallurgy. Aluminium is obtained on a large scale by electrolysing alumina dissolved in a molten electrolyte, with graphite anodes by the Bayer–Hall–Héroult process. The overall reaction is

$$2Al_2O_3 + 3C \rightarrow 4Al + 3CO_2$$

or

$$2Al_2O_3 + 6C \rightarrow 4Al + 6CO$$

The oxygen discharged at the anode combines with the carbon electrode. The theoretical decomposition voltage for this reaction is only about 1.2 V, since much of the energy required to decompose the alumina is supplied by the oxidation of the carbon.

Aluminium oxide melts at 2000°C, and the technical difficulties in working at this temperature would be almost insuperable. The oxide dissolves in cryolite (Na_3AlF_6) and the eutectic contains about 10 percent alumina and melts below 1000°C. Magnesium fluoride and up to 5 percent calcium fluoride are common additions; the working temperature is then about 950°C. Molten aluminium is collected as the lower layer to prevent oxidation. It is not economic to prepare crude aluminium and then purify it, so very pure alumina is used in the process, and the anodes should also be pure and ash-free. The product is aluminium of about 99.7 percent purity, with traces of iron and silicon. Even purer metal can be obtained from this by **electrorefining** in a bath similar to that used in the preparation.

The cells consist of iron containers with a carbon lining, which serve as the cathode. Molten aluminium collects on the floor of the cell. Above this is the fused electrolyte, into which the carbon anodes dip, with a crust of solidified electrolyte at its surface.

The exact composition of the electrolyte is unknown; it could contain a large number of constituents (e.g., AlO_2^-, AlF_6^{3-} and other oxygen and fluorine compounds of aluminium, as well as sodium, aluminium and fluoride ions). Consequently, there is uncertainty about the primary cathodic and anodic reactions. The simplest cathodic reaction would be

$$Al_2O_3 + 3e \rightarrow Al + AlO_3^{3-}$$

At the anode the discharge of fluoride, or complex ions containing it, is unlikely, so the probable reaction involves AlO_2^- or AlO_3^{3-}: for example

$$2AlO_2^- \rightarrow Al_2O_3 + O + 2e$$

followed by

$$2O + C \rightarrow CO_2$$

The current efficiency is about 80 percent, the loss being due to the metal cloud that leaves the cathode.

A modern development has been made by Alcona in which the melt $AlCl_3$–$LiCl$–$NaCl$ is electrolysed between graphite electrodes. Aluminium chloride is regenerated by reaction between chlorine liberated at the anode and coke and alumina. Bipolar cells give an improved use of cell volume and the current efficiency is greater than that of the Hall–Héroult process. See Kirk-Othmer (1981), Pletcher (1982).

Amalgam electrode. A variation of a metal–metal ion electrode in which the metal is present as an amalgam (i.e. dissolved in mercury) rather than in the pure form; a platinum wire is used to make electrical contact. The reaction occurring is the same as that at a metal electrode; the mercury plays no chemical role. In general, the potentials of amalgam electrodes are more reproducible than those of solid metal electrodes, which are sensitive to surface impurities and mechanical strains in the crystalline solid. The e.m.f.

12 Ammonia-sensing probe

of a cell incorporating amalgam electrodes depends on the activity of the metal in the amalgam (*see* **Concentration cell**).

Amalgam electrodes allow active metals (e.g., sodium and potassium) to be used as electrodes. Reproducible, reversible potentials can be obtained at very low concentrations of the metal. It is best to use dropping electrodes, in which the amalgam drips slowly into the electrolyte, so the surface is always fresh. The potential of the amalgam is compared with that of the pure metal by using both in a cell in which the electrolyte is dissolved in a solvent that does not attack the pure metal (e.g., ethylamine). Thus to determine $E^{\ominus}_{K^+,K}$, the sum of the standard e.m.f. values of the two cells

$$Hg,K \mid KCl \mid AgCl, Ag \quad \text{and} \quad K \mid KI \text{ in ethylamine} \mid K, Hg$$

gives the standard e.m.f. of the hypothetical cell

$$K \mid KCl \mid AgCl, Ag$$

as

$$E^{\ominus} = 3.1464 \text{ V}$$

Hence

$$E^{\ominus}_{K^+,K} = 3.1464 - 0.2224 = 2.9240 \text{ V}$$

See Ives and Janz (1961).

Ammonia-sensing probe. The most widely used **gas-sensing membrane probe**. Such probes are employed in the analysis of fresh water, effluent and sewage. When an ammonium-sensing probe is in contact with a solution containing gaseous ammonia, the internal solution between the pH electrode and the gas-permeable membrane gains or loses ammonia gas through the latter, until the partial pressure (activity) of ammonia is the same on both sides. The pH of the internal solution depends on the concentration of ammonia according to a Nernstian equation (*see* **Nernst equation**); the response range is from 1 to 10^{-7} mol dm^{-3}. Although the probe is capable of detecting low levels of ammonia, it cannot detect very small changes in the ammonia concentration at high concentrations. The potential of the probe is affected by change of temperature, hence samples and standards should be at the same temperature (in the range 5–40°C).

The solution under test must be at pH > 12 to ensure that the ammonia is in the free state in solution; any ammonia present as a complex must be released (e.g., with EDTA to destroy the complex). Volatile species (e.g., organic amines) which are basic in solution interfere with the measurement.

A sensor based on immobilized nitrifying bacteria and an oxygen electrode has been described for the amperometric determination of ammonia. One genus of nitrifying bacteria, *Nitrosomonas* sp., utilizes ammonia as the sole source of energy, and oxygen is consumed by the respiration

$$NH_3 + 3/2 O_2 \xrightarrow{\text{Nitrosomonas sp.}} NO_2^- + H_2O + H^+$$

while *Nitrobacter* sp. oxidizes nitrite to nitrate

$$NO_2^- + 1/2 O_2 \xrightarrow{\text{Nitrobacter sp.}} NO_3^-$$

The oxidation of both nitrite and ammonia proceeds at a high rate, and the oxygen consumed is determined by an **oxygen probe**. The pH of the sample solution must be greater than the pK of ammonia (9.25).

Further reading
T. Okada, I. Karube and S. Suzuki, Ammonium ion sensor based on immobilized nitrifying bacteria and a cation-exchange membrane, *Anal. Chim. Acta*, **135**, 159 (1982).

Ampere (A; dimensions: $\varepsilon^{1/2} \, m^{1/2} \, l^{3/2} \, t^{-2}$). A basic SI unit; defined as that **current** which, if maintained in two parallel conductors of infinite length, of negligible cross-section and placed 1 metre apart in a vacuum, would produce a force between the conductors equal to 2×10^{-7} newton per metre of length.

The international ampere is the current which deposits 1.118 00 mg of silver per second from a standard solution of silver nitrate. *See also* **Electric units**.

Amperometric titrations. *See* **Electrometric titrations**.

Anion. A negative ion (e.g., Cl^-). When a current is passed through an electrolyte solution, the anions move towards the positive electrode, the **anode**. *See also* **Cation**.

Anode. The electrode where oxidation (i.e. loss of electrons: red → ox + ne) occurs; that where reduction occurs is the **cathode**. In an electrochemical **cell** acting as a source of electricity (i.e. chemical energy being converted into electrical energy), the anode is at a lower potential than the cathode. For the reaction to proceed the species undergoing reduction withdraws electrons from its electrode (cathode) leaving a net positive charge on it. The opposite occurs at the other electrode (anode) where the species undergoing oxidation donates electrons to the electrode thereby giving it a net negative charge.

When the cell is used in electrolysis (i.e. electrical energy being converted into chemical energy), the anode is still the electrode where oxidation occurs but now electrical work is being used to produce a chemical change. Anions must be induced to move to the anode for oxidation, while cations have to be

14 Anodic stripping voltammetry

drawn to the cathode for reduction. The anode must, therefore, be relatively more positive than the cathode. In summary

	anode	cathode
process	oxidation red → ox + ne	reduction ox + ne → red
potential		
galvanic cell	low (−)	high (+)
electrolytic cell	high (+)	low (−)

Anodic stripping voltammetry. *See* **Stripping voltammetry.**

Anodizing. Formation of an oxide or chloride film or coating on certain metals by electrolysis in a suitable solution. On the application of an electric potential to a cell in which the metal to be treated is the anode, the oxidizing conditions convert the surface of the metal to the oxide (or chloride). The oxide (or chloride) film is essentially an integral part of the metal. Among the properties which may be altered by anodizing are resistance to corrosion and abrasion, hardness, appearance, and reflection and radiation characteristics. *See also* **Corrosion; Silver–silver chloride electrode.**

Anolyte. Electrolyte surrounding the **anode** in an electrochemical cell.

Antimony–antimony oxide electrode. An electrode consisting of pure antimony metal dipping into a solution of an electrolyte. A skin of antimony oxide forms on the surface of the metal, and this is in equilibrium with the antimony ions in solution

$$Sb_2O_3 + 6H^+ + 6e \rightleftharpoons 2Sb^{3+} + 3H_2O$$

The potential of the electrode is given by an equation of the form

$$E_{Sb^{3+},Sb} = E' + \frac{RT}{F} \ln a_{H^+}$$

where E' is a constant, which must be determined experimentally for each electrode. The electrode is very robust and can be used to determine the pH of a solution in the pH range 4–12 with an accuracy of ±0.2 pH unit. Although it does not contaminate the liquid under test, it cannot be used in the presence of dissolved oxygen, oxidizing agents, hydrogen sulphide or heavy metal ions, or in highly acid or alkaline solutions. *See also* **Half cell; Oxide electrode.**

Arrhenius electrolytic dissociation theory. Arrhenius (1882) was the first to advance the view that an electrolyte, when dissolved in water, dissociates extensively into free ions. His theory soon found support in the work of van't Hoff on the colligative properties of dilute solutions, since it was found that a salt such as sodium chloride had almost twice and calcium chloride nearly three times the effect on the vapour pressure, etc. of a solvent as did a normal (undissociated) solute.

An electrolyte MA was thus supposed to exist in solution as an ionized fraction (α) of free ions M^+ and A^- in equilibrium with a fraction $(1 - \alpha)$ of undissociated MA molecules. At infinite dilution dissociation is complete—a necessary consequence of the law of mass action—and Arrhenius therefore proposed to determine α from the relationship $\alpha = \Lambda/\Lambda^\circ$. On this basis, the **dissociation constant** of an electrolyte could be calculated from conductance measurements

$$K = \frac{c_{M^+} c_{A^-}}{c_{MA}} = \frac{\alpha^2 c^2}{(1-\alpha)c} = \frac{\Lambda^2 c}{\Lambda^\circ (\Lambda^\circ - \Lambda)}$$

This equation is often referred to as Ostwald's dilution law. It depends on the assumptions that: (1) the **molar ionic conductivity** values are constant and independent of the concentration of the electrolyte; and (2) the ions in dilute solution behave as ideal solutes. Both of these assumptions have proved to be incorrect and have been corrected in the interionic theory of Debye, Hückel and Onsager. *See also* **Conductance equations; Conductance of aqueous solutions**.

Auxiliary electrode. *See* **Three-electrode system.**

B

Battery. *See* **Electrochemical storage.**

Beryllium electrometallurgy. Beryllium is obtained by the electrolysis of a fused mixture of the equimolar eutectic of sodium and beryllium chlorides (mp 210°C). The anode is graphite and the cathode is the nickel vessel in which the process is conducted. The operating temperature is about 350°C. At this temperature, beryllium is solid and is removed from the cell from time to time. Chlorine is evolved at the anode. The **current efficiency** of the process is about 50 percent.

Biocatalytic membrane electrode. *See* **Biosensor.**

Biochemical standard state. In physical chemistry the standard state in solution refers to the situation in which all the reactants and products are at unit activity. In biochemistry the same convention is adopted, except that activities are replaced by concentrations and the hydrogen ion concentration in the standard state is 10^{-7} mol dm^{-3} (i.e. pH 7.0). In biochemical processes $\Delta G^{\ominus\prime}$ replaces $\Delta G^{\ominus}$ and $E^{\ominus\prime}$ replaces $E^{\ominus}$.

For the process

$$A + B \rightarrow C + xH^+$$

the van't Hoff isotherm (using concentrations instead of activities) can be written

$$\Delta G = \Delta G^{\ominus} + RT \ln \frac{(c_C/c_C^{\ominus})(c_H^{\ominus}/c_{H^+}^{\ominus})^x}{(c_A/c_A^{\ominus})(c_B/c_B^{\ominus})}$$

where $c_A^{\ominus}$, $c_B^{\ominus}$, $c_C^{\ominus}$ and $c_{H^+}^{\ominus}$ are the standard state concentrations. Inserting the appropriate values of these gives

$$\Delta G = \Delta G^{\ominus} + RT \ln \frac{(c_C c_{H^+})^x}{c_A c_B}$$

$$= \Delta G^{\ominus\prime} + RT \ln \frac{c_C (c_{H^+}/10^{-7})^x}{c_A c_B}$$

whence it follows that

and
$$\Delta G^{\ominus} = \Delta G^{\ominus\prime} + xRT \ln \frac{1}{10^{-7}}$$

$$E^{\ominus} = E^{\ominus\prime} - \frac{xRT}{nF} \ln \frac{1}{10^{-7}}$$

when $T = 298$ K and $x = 1$

$$\Delta G^{\ominus} = \Delta G^{\ominus\prime} + 40.0$$

and

$$E^{\ominus} = E^{\ominus\prime} - \frac{0.414}{n}$$

where $\Delta G^{\ominus}$ has the units of kJ mol^{-1} and $E^{\ominus}$ the units of V. Thus a reaction releasing hydrogen ions is more spontaneous at pH 7 than at pH 0; the converse is true for a reaction in which hydrogen ions are one of the reactants. For reactions not involving hydrogen ions, $\Delta G^{\ominus} = \Delta G^{\ominus\prime}$ and $E^{\ominus} = E^{\ominus\prime}$.

Bioelectrochemistry. *See* **Biological oxidation–reduction systems; Biosensor; Electron transport chain; Ion-selective electrode; Ion-selective microelectrode; Microbial fuel cell; Mini-ion-selective electrode; Polarography.**

Further reading
M. J. Allen and P. N. R. Usherwood (ed.), Charge and field effects in biosystems, in *Proceedings of the 1983 International Symposium on Bioelectrochemistry and Bioenergetics*, Abacus Press, Tunbridge Wells (1984)
M. Blank (ed.), Bioelectrochemistry, ions, surfaces, membranes, in *Advances in Chemistry Series*, no. 188, Washington, American Chemical Society (1980)
J. O'M. Bockris *et al.* (ed.), *Comprehensive Treatise of Electrochemistry*, vol. 10, Plenum Press, New York (1984)

Bio-fuel cell. *See* **Microbial fuel cell.**

Biological oxidation–reduction systems. As with inorganic systems, electrons are transferred in biological redox chains in the direction of increasing electrode potential. The redox potential is given by the **Nernst equation**

$$E = E^{\ominus\prime} + \frac{RT}{nF} \ln \frac{c_{\text{ox}}}{c_{\text{red}}}$$

in which activities are replaced by concentrations. The standard redox (reduction) potential $E^{\ominus\prime}$ of a biological couple is based on pH 7 rather than pH 0 (*see* **Biochemical standard state**), thus at 298 K

$$E^\ominus = E^{\ominus\prime} \pm \frac{0.414}{n}$$

where the negative (positive) sign refers to reactions in which hydrogen ions are one of the products (reactants). $E^{\ominus\prime}$ depends on the pH of measurement

$$\frac{dE^{\ominus\prime}}{dpH} = \frac{0.059x}{n}$$

where x is the number of hydrogen ions participating in the redox reaction. Biological half reactions with high $E^{\ominus\prime}$ values are good oxidizing agents.

Substituents and ligands with a negative charge decrease the redox potential of proteins. Steric factors and changes of spin state have a complex influence on the potentials of metalloenzymes.

The standard redox potential can be determined by measuring the potential of a platinum electrode (against a reference electrode) dipping into solutions of different values for c_{ox}/c_{red} and determining the intercept of the plot of potential against ln (c_{ox}/c_{red}) (see **Redox electrode systems**). Approximate values may be obtained from colorimetric measurements using a redox indicator.

Tabulated values of $E^{\ominus\prime}$ (see Appendix, Table 6) are used to calculate standard free energy changes during the transfer of electrons. The complete oxidation of $NADH_2^+$ to NAD^+ by molecular oxygen occurs by a series of stages in the **electron transport chain**, which can be represented by

$$NAD^+ + 2H^+ + 2e \rightleftharpoons NADH_2^+ \qquad E^{\ominus\prime} = -0.32 \text{ V}$$
$$\tfrac{1}{2}O_2 + 2H^+ + 2e \rightleftharpoons H_2O \qquad E^{\ominus\prime} = +0.82 \text{ V}$$

$$NADH_2^+ + \tfrac{1}{2}O_2 \rightleftharpoons NAD^+ + H_2O$$

for which the standard e.m.f. of a cell in which these reactions occur at the two electrodes is $0.82 - (-0.32) = 1.14$ V. The standard free energy change is

$$\Delta G^{\ominus\prime} = -nFE^{\ominus\prime} = -2 \times 1.14 \times 96\,500 = -220 \text{ kJ mol}^{-1}$$

Further reading
G. Dryhurst, K. M. Kadish, F. Scheller and R. Renneberg, *Biological Electrochemistry,* Academic Press, New York (1982)

Bipolar electrode. An electrode unconnected with the electricity supply which is placed between the anode and cathode in a cell. Under the influence of the electric field, anions travel to the bipolar electrode and are discharged, and the liberated electrons traverse the plate and initiate a cathodic reaction at the opposite surface. Compared with a series of separate cells (*see* Figure 7), the system saves engineering and eliminates contact resistances. A problem that arises with bipolar cells is the possibility of electrical leakage currents, also

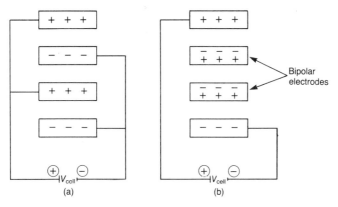

Figure 7. Electrical connections in (a) a series of monopolar cells and (b) a bipolar cell

called shunt or bypass currents, between the electrodes in neighbouring cells, if there is not an effective barrier between cells. A leakage current can lead to a loss in efficiency, corrosion and safety hazards (e.g., the mixing of hydrogen and oxygen in a water electrolysis cell).

Biosensor. Electrode developed for environmental or clinical analysis and fermentation control. A biosensor comprises a chemically responsive material (enzyme, lectin, antibody, microorganism or organelle) immobilized in close proximity to a suitable transducing element designed to convert a chemical response into an electrical response. The transducer may take one of a number of forms including a metal or semiconductor electrode, an **ion-selective electrode**, a transistor, a piezoelectric crystal or an optoelectronic device. The transducer must be highly specific for the compound of interest and responsive to the range required (e.g., in a biological sample), have a fast response time, be capable of miniaturization, in some applications be amenable to sterilization, be robust and be reliable.

Potentiometric sensors (ion-selective, glass, solid-state and gas-sensing electrodes) are usually chosen to detect the product of the biological (e.g., enzymatic) reaction (*see* Table 2). A **gas-sensing membrane probe** is most frequently used because of its high degree of selectivity. Such probes suffer, however, from long response and recovery times. Amperometric sensors measure the flux of an electroactive species which they are designed to detect. They may be used to measure a decrease in the concentration of one reactant (e.g., oxygen) or the increase in concentration of a product (e.g., hydrogen peroxide) (*see* Table 3).

Enzymes may be immobilized by: (1) crosslinking to serum albumin, Teflon or nylon using glutaraldehyde as a bifunctional reagent; (2) occlusion in a polymer; and (3) liquid trapping. Alternatively, the biocatalyst may be

Table 2
Potentiometric sensors for immobilized enzymes

Sensor	Species detected	Typical substrates
H^+ glass electrode	H^+	Penicillin, glucose, urea, acetylcholine
NH_4^+ glass electrode	NH_4^+	Urea, D- and L-amino acids, including glutamic acid, glutamine
NH_3 gas sensor	NH_3	Asparagine, creatinine, 5'-AMP, urea, nitrilotriacetic acid,[a] serine[a]
CO_2 gas sensor	CO_2	Urea, uric acid, tyrosine, glutamic acid[a,b]
I^- solid-state electrode	I^-	Glucose
CN^- solid-state electrode	CN^-	Amygdalin

[a] Intact bacterial cells rather than enzymes.
[b] Tissue slices rather than enzymes.

retained at the electrode surface by a polymer membrane (with appropriate molecular weight cut-off) which allows diffusion of the substrate into the biocatalytic area and, at the same time, prevents diffusion of the biocatalyst away from the electrode surface. With tissue slices, a cellophane dialysis membrane between the slice and the gas-sensing probe protects the sensor from lipids, proteins and other biological material (see Figure 8).

Ion-selective electrodes have been successfully developed for monitoring blood sodium, potassium, lithium and calcium ions and for a variety of drugs including novocaine, codeine and morphine. The marriage of specific enzymes with ion- or gas-sensing electrodes has led to the development of enzyme electrodes and enzyme immunoelectrodes to give a simple and

Table 3
Amperometric sensors for immobilized enzymes

Sensor	Electroactive species detected	Typical substrates
Pt, Clark oxygen electrode	O_2	L-Amino acids, glucose, alcohols, aldehydes, sucrose
Pt	H_2O_2	L-Amino acids, glucose
Pt	I_2	Glucose
Pt	$Fe(CN)_6^{4-}$	Lactic acid

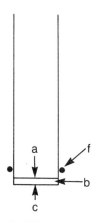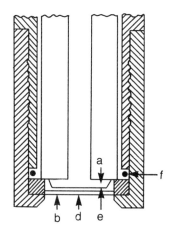

Figure 8. Different types of potentiometric enzyme electrodes: a, Potentiometric sensor; b, trapped or immobilized enzyme; c, nylon spacer; d, dialysis membrane; e, gas-permeable membrane; f, O-ring seal

reproducible electrochemical procedure for the assay of hormones, metabolites and proteins. Penicillinase from *Bacillus cereus* has been used as an enzyme label for the estimation of testosterone levels (1–10^{-3} ng dm^{-3}) with a conventional glass electrode as the transducer.

Although immobilized enzyme electrodes are highly selective, enzymes tend to be expensive and the enzyme reaction may be inhibited by compounds present in the sample (e.g., fermentation media, broths). To overcome these problems microbial sensors have been developed, in which living microorganisms are immobilized on oxygen electrodes. When inserted in a sample solution containing an organic compound, this compound diffuses to the bacteria where it is assimilated. As a result of this, the respiratory activity (i.e. oxygen uptake) of the cells increases with a corresponding decrease in the oxygen diffusing to the probe. This decrease is monitored amperometrically.

An opticoelectronic sensor, designed for the estimation of proteins, depends on the change of colour of a dye on binding to a protein. The binding of bromocresol green to human serum albumin at pH 3.8 is accompanied by a change in colour from yellow to green or blue. The dye is covalently attached to a transparent cellophane membrane, sandwiched between a red light-emitting diode (LED) and a silicon photodiode with integral amplifier. When the protein solution flows past the bound dye, there is a change in the colour which produces a change in the output voltage of the diode. The sensor responds to human serum albumin concentrations in the range 0–45 mg cm^{-3} (the output is linear with concentration over the range 5–35 mg cm^{-3}). The effect of albumin on the bound dye is completely reversible, and the assembly can be included in a continuous flow system.

Bjerrum's ion association theory

Further reading

P. W. Carr and L. D. Bowers, *Immobilized Enzymes in Analytical and Clinical Chemistry*, John Wiley, New York, p. 197 (1980)

I. Karube and S. Suzuki, Microbial sensor for fermentation control, in *Advances in Biotechnology*, Marcel Dekker, New York, p. 355 (1982)

S. Suzuki and I. Karube, Bioelectrochemical sensor with use of immobilized enzymes, whole cells and proteins, in *Applied Biochemistry and Bioengineering*, vol. 3, Academic Press, New York (1981)

P. Vadgama, Enzyme electrodes, in *Ion-Selective Electrode Methodology*, vol. II, A. K. Covington (ed.), CRC Press, Cleveland, p. 23 (1979)

Bjerrum's ion association theory. *See* **Ion pair.**

Breathalyser. *See* **Alcohol meter.**

Brønsted–Bjerrum equation. *See* **Ion pair.**

Buffer solution. A solution that maintains a nearly constant pH despite the addition of small amounts of acid or alkali. Thus a solution of sodium chloride in water (pH ≈ 7) cannot maintain its pH as hydrogen or hydroxide ions are added. Many solutions, however, have a considerable reserve to remove added hydrogen or hydroxide ions and hence maintain a constant pH; such systems are said to show buffer action.

Most buffer solutions consist of one or more weak acids and their conjugate bases; buffer solutions containing a weak base and conjugate acid (e.g., ammonium hydroxide and ammonium chloride) are less common. For the solution of a weak acid and its conjugate base in water

$$NaA \rightarrow Na^+ + A^-$$
$$H_2O + HA \rightleftharpoons H_3O^+ + A^-$$

any added hydrogen ion is removed by the reverse reaction to give undissociated acid, whereas added hydroxide ion is removed by the free acid

$$HA + OH^- \rightarrow A^- + H_2O$$

Continual addition of acid or base eventually results in complete displacement of equilibrium one way, and buffer action ceases. From the definition of **dissociation constant**, it follows that

$$pH = pK_a + \log \frac{c_{A^-}}{c_{HA}} + \log \frac{\gamma_{A^-}}{\gamma_{HA}}$$

$$= pK_a + \log \frac{c_{A^-}}{c_{HA}} - AI^{1/2} + CI$$

where c_{A^-} and c_{HA} are the total concentrations of salt and free acid, respectively, and A and C are the Debye–Hückel constants. The simplified version of this equation, without the activity correction terms—generally known as the Henderson equation—shows that the pH of a buffer solution is determined by (1) the pK_a of the weak acid and (2) the ratio c_{A^-}/c_{HA}. It gives no information, however, about the buffer capacity or the concentrations of acid and salt to use.

Buffer capacity (β) is measured by dB/dpH, where dB is the amount of strong base which when added to a buffer solution produces an increment dpH in the pH (i.e. the reciprocal of the slope of the pH-titration curve at a given point). In the effective buffer region (neglecting activity coefficients)

$$\beta = \frac{2.3 a K_a c_{H^+}}{(K_a + c_{H^+})^2}$$

that is β is directly proportional to a, the total concentration of free acid and salt. β is maximal when $K_a = c_{H^+}$, that is when pH = pK_a and $c_{A^-}/c_{HA} = 1$; this is the mid-point of the neutralization curve (*see* Figure 9). β_{max} (= $2.3a/4$) is independent of K_a.

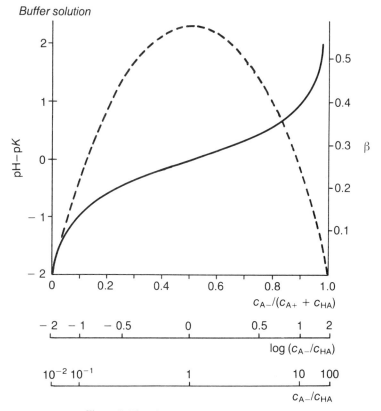

Figure 9. Titration curve and buffer capacity

24 Buffer solution

If c_{H^+}/c_{HA} is increased or decreased by a factor of ten (i.e. $pH = pK_a \pm 1$), then at these extreme values $\beta = 2.3a/12$, which although appreciable is about $\frac{1}{3}\beta_{max}$. Beyond these extremes, β decreases so rapidly that there is no buffer action.

For a given weak acid, the buffer solution is only of use for 1 pH unit on either side of the pK_a, hence it is necessary to select an acid of pK_a nearest the required pH. It is not necessary for the acid or salt to be a neutral molecule: for example, in Sørensen's phosphate buffer solution

$$H_2PO_4^- + H_2O \rightleftharpoons HPO_4^{2-} + H_3O^+$$
$$(NaH_2PO_4:acid) \qquad (NaHPO_4:salt)$$

Then c_{A^-}/c_{HA} is calculated from the Henderson equation, and the buffer solution of this ratio is then prepared. The effect of variation of ionic strength on the pH can be calcuated from

$$pH = pK_a + \log \frac{c_{A^-}}{c_{HA}} - (2n - 1)AI^{1/2} + CI$$

which applies to the nth stage of neutralization of a polyprotic acid. Hence

$$\frac{dpH}{dI^{1/2}} = -(2n - 1)A + CI^{1/2}$$

Thus the pH may increase or decrease with I, depending on conditions.

The data for many buffer solutions have been tabulated (see Table 4); the pH values only apply to the concentrations quoted. Mixed buffer solutions, providing good buffer capacity over a wide range of pH values, can be obtained by using mixtures of weak acids (e.g., McIlvaine's buffer (pH 2.2–8.0), consisting of disodium hydrogen phosphate and citric acid, has five conjugate acid–base pairs).

Table 4
Buffer solutions

Solutions	pH range
HCl and KCl	1.0–2.2
Glycine and HCl	1.0–3.7
KH phthalate and HCl	2.2–3.8
Acetic acid and NaOH	3.7–5.6
KH_2PO_4 and NaOH	5.8–8.0
Boric acid and borax	6.8–9.2
Diethylbarbituric acid and sodium salt	7.0–9.2
Boric acid and NaOH	7.8–10.0
Na_2HPO_4 and NaOH	11.0–12.0

Butler–Volmer equation

Although solutions of strong acids and bases are not normally classified as buffer solutions, they have large buffer capacity at high concentrations. For a strong acid or base

$$\beta = 2.3(c_{H+} + c_{OH-})$$

Thus at pH 1 or 13, $\beta = 0.23$ in agreement with the relatively flat pH–neutralization curves in the early stages.

Butler–Volmer equation. Consider a simple one-electron redox reaction

$$ox^z + e \rightarrow red^{z-1}$$

For example

$$Fe^{3+} + e \rightarrow Fe^{2+} \quad \text{or} \quad Ce^{4+} + e \rightarrow Ce^{3+}$$

The free energy of the system may vary with reaction coordinate as shown in Figure 10. The series of small activation barriers A—B and C—D represents the progress of the oxidized species to the outer Helmholtz plane and the movement away from the electrode of the product. As Figure 10 is drawn the rate-determining step is the electron transfer B—C via some activated complex $\ddagger$. The object of the exercise is to derive an equation which relates the rate of the electron transfer to the potential at the electrode.

By absolute reaction rate theory the rate of the forward electron transfer is

$$v_f = a_{ox} \left(\frac{\kappa_f kT}{h} \right) \exp \left(-\frac{\Delta \bar{G}_f^{\ddagger}}{RT} \right) \tag{1}$$

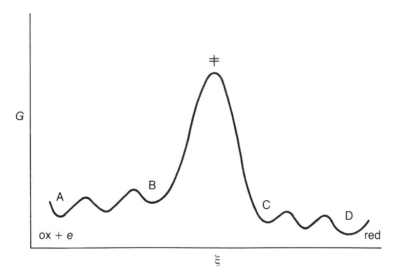

Figure 10. Free energy of system along reaction coordinate of an electrochemical reaction

Butler–Volmer equation

a_{ox} is the activity of reactant ox^z at the outer Helmholtz plane where the reaction is assumed to take place. κ_f is the forward transmission coefficient and $\Delta\bar{G}_f^{\ddagger}$ is the forward free energy barrier over which the system must pass. The free energy of the initial and final states may be written in terms of potential-independent and potential-dependent terms: for example, the free energy of the initial state $(\bar{G}_B)$, $ox^z + e$ is

$$\bar{G}_B = \mu_{ox} + \mu_e + zF\phi_p - F\phi_m \tag{2}$$

μ_{ox} and μ_e are the chemical potentials of ox^z and electron, respectively, ϕ_p and ϕ_m are the inner potentials (*see* **Interfacial potential**) at the outer Helmholtz plane (or pre-electrode state as it is more correctly known) and in the metal. The introduction of ϕ_p allows double layer effects to be taken into account. In the derivation given here it will be assumed (as is often the case) that $\phi_p = 0$. If the potential at the outer Helmholtz plane is zero, therefore an electron travelling across from metal to ox^z must fall through the total electrode potential. At some stage the activated complex is reached and thus only a fraction $(0 > \beta > 1)$ of the applied potential will contribute to the energy of the activated complex. The forward activation energy is

$$\Delta\bar{G}_f^{\ddagger} = \bar{G}^{\ddagger} - \bar{G}_B = \Delta G_f^{\ddagger} + \beta F\phi_m \tag{3}$$

$\Delta G_f^{\ddagger}$ consists of all the terms that are independent of potential and is the activation free energy in the absence of potential. β is the **symmetry coefficient** and usually has a value near 0.5. The backward activation energy is written in an analogous manner except that the contribution of the potential to the increase in activation energy is now $(1-\beta)F\phi_m$.

$$\Delta\bar{G}_b^{\ddagger} = \Delta G_b^{\ddagger} - (1-\beta)F\phi_m \tag{4}$$

The respective forward and backward rates are

$$v_f = \lambda a_{ox}\left(\frac{\kappa_f kT}{h}\right)\exp\left(-\frac{\Delta G_f^{\ddagger}}{RT}\right)\exp\left(-\frac{\beta F}{RT}\phi_m\right) \tag{5}$$

$$v_b = \lambda a_{red}\left(\frac{\kappa_b kT}{h}\right)\exp\left(-\frac{\Delta G_b^{\ddagger}}{RT}\right)\exp\left(\frac{(1-\beta)F}{RT}\phi_m\right) \tag{6}$$

where λ is the thickness of the reaction layer. The current density is related to the rate of reaction by **Faraday's laws**.

$$i_f = Fv_f \quad \text{and} \quad i_b = Fv_b$$

and the net forward current $i = i_f - i_b$. The value of ϕ_m is not measurable but the **overpotential** (η), that is the difference between the potential when current flows and at equilibrium, may be measured

$$\phi_m = \phi_{m,e} + \eta \tag{7}$$

Therefore

$$i_f = F\lambda a_{ox}\left(\frac{\kappa_f kT}{h}\right)\exp\left(-\frac{\Delta G_f^{\ddagger}}{RT}\right)\exp\left(-\frac{\beta F}{RT}\phi_{m,e}\right)\exp\left(-\frac{\beta F}{RT}\eta\right) \quad (8)$$

$$i_b = F\lambda a_{red}\left(\frac{\kappa_b kT}{h}\right)\exp\left(-\frac{\Delta G_b^{\ddagger}}{RT}\right)\exp\left[\frac{(1-\beta)F}{RT}\phi_{m,e}\right]\exp\left[\frac{(1-\beta)F}{RT}\eta\right] \quad (9)$$

Note that for a reduction η is negative. At equilibrium, when no net current flows, $i_f = i_b = i_o$ and $\eta = 0$. i_o is the **exchange current density** and from Eq. (8)

$$i_o = F\lambda a_{ox}\left(\frac{\kappa_f kT}{h}\right)\exp\left(-\frac{\Delta G_f^{\ddagger}}{RT}\right)\exp\left(-\frac{\beta F}{RT}\phi_{m,e}\right) \quad (10)$$

$$= F\lambda a_{red}\left(\frac{\kappa_b kT}{h}\right)\exp\left(-\frac{\Delta G_b^{\ddagger}}{RT}\right)\exp\left[\frac{(1-\beta)F}{RT}\phi_{m,e}\right] \quad (11)$$

Combining Eqs. (8), (9), (10) and (11)

$$i = i_f - i_b = i_o\left\{\exp\left(-\frac{\beta F}{RT}\eta\right) - \exp\left[\frac{(1-\beta)F}{RT}\eta\right]\right\} \quad (12)$$

This is the Butler–Volmer equation. The form of the i versus η curve is shown in Figure 11 for different values of β. Two approximations may be made to the Butler–Volmer equation at high overpotential (*see* **Tafel equation**) and at low overpotential. Near equilibrium (when $\eta < 0.01$ V) a linear approximation may be made to the exponential terms in Eq. (12)

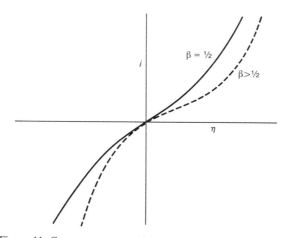

Figure 11. Current–overpotential curves for different values of β

28 Button cell

$$i \approx i_o\left(1 - \frac{\beta F}{RT}\eta\right) - i_o\left[1 + \frac{(1-\beta)F}{RT}\eta\right] \quad (13)$$

$$i \approx -\frac{i_o F}{RT}\eta \quad (14)$$

The current follows Ohm's law with $RT/Fi_o A$ having the units of resistance, where A is the area of the electrode. The above equations assume that the rate-limiting step is electron transfer. If diffusion significantly affects the rate the equations need to be modified (*see* **Diffusion-limited current**). The form of the current–overpotential relation for multistep reactions is given under **electrode reaction mechanisms**. See Bockris and Reddy (1973).

Button cell. A flat cell in which anode and cathode discs are pressed together on either side of a separator. The design is simple with low-cost, easily fabricated components. Because of the low resistance to pressure and inefficient operation of this design, button cells are limited in size and application to low discharge rates.

Most dry batteries may be made as button cells. Figure 12 shows a typical design of a nickel–cadmium button cell.

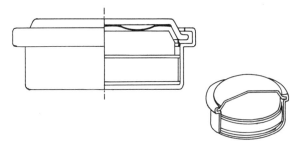

CUTAWAY VIEWS OF STANDARD RATE BUTTON CELL

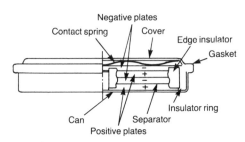

CROSS SECTION VIEW—DOUBLE PLATE MOULDED ELECTRODE
HIGH RATE BUTTON CELL

Figure 12. Eveready sealed nickel–cadmium button cells. From T. R. Crompton, *Small Batteries*, vol. 1, *Secondary Cells*, Macmillan, London, p. 53 (1982)

C

Cadmium electrometallurgy. Cadmium is obtained by the electrolysis of cadmium sulphate solution, the process being very similar to that employed for zinc (*see* **Zinc electrometallurgy**). Free acid is formed during the electrolysis, and this is used in a recycling process to take more cadmium into solution. The main source of cadmium is from the purification of zinc sulphate solution prior to its electrolysis; the cadmium is precipitated from this by the addition of zinc metal at 70°C whence spongy cadmium is formed.

Calcium electrometallurgy. Calcium is prepared by electrolysing fused calcium chloride at about 800°C; some calcium fluoride may be added to lower the melting point. Chlorine is produced and led off from the graphite anodes, whereas molten calcium is discharged at the metal cathode. It has a considerable solubility in the electrolyte, so a water-cooled cathode is used and the column of solidified calcium is continuously withdrawn from the electrolyte as electrolysis proceeds. An iron cathode drawn up from the melt has solid calcium (mp 845°C) covered with calcium chloride (mp 772°C) and thus the calcium is protected from the oxidizing atmosphere.

Calomel electrode. A reference electrode that consists essentially of mercury, mercury (I) chloride and potassium chloride solution of specified concentration (*see* Figure 13) (i.e. Hg, $Hg_2Cl_2 | KCl(aq)$). The electrode potential is given by

$$E_{cal} = E^{\ominus}_{cal} + \frac{RT}{2F} \ln K_s - \frac{RT}{F} \ln a_{Cl^-}$$

$$= E' - \frac{RT}{F} \ln a_{Cl^-}$$

where

$$K_s = a_{Hg^{2+}_2} a_{Cl^-}^2$$

The electrode thus behaves like a reversible chlorine electrode; the electrode potential (E) depends on the concentration of the potassium chloride solution and the temperature T/K (*see* Table 5).

30 Calomel electrode

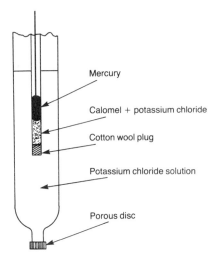

Figure 13. Calomel electrode

The first electrode listed in Table 5 is preferred for accurate work as it has the lowest temperature coefficient; the saturated electrode is the most convenient owing to the ease of replacement of the solution. In commercial electrodes, the liquid junction between the half cell and the test solution is made by leakage of the potassium chloride solution through a ceramic disc.

The aqueous calomel electrode is the most common reference electrode used in non-aqueous systems; connection to the non-aqueous solution is made via a salt bridge in which the junction is made through a sintered disc. Although the junction involves the interdiffusion of different solvents, the liquid junction potential is sufficiently constant for electrode potentials in non-aqueous solvents to be measured.

Half cells involving glacial acetic acid have been described

$$\text{Hg}, \text{Hg}_2\text{Cl}_2 \mid \text{NaCl}, \text{NaClO}_4 \text{ saturated solution of both salts in acetic acid} \mid \ldots$$

Table 5

KCl concentration	E/V
0.1 mol dm^{-3}	$0.3335 - 7.0 \times 10^{-5}(T - 298)$
1.0 mol dm^{-3}	$0.2810 - 2.4 \times 10^{-4}(T - 298)$
Saturated	$0.2420 - 7.6 \times 10^{-4}(T - 298)$

Since the solubility of sodium chloride is so small in acetic acid, sodium perchlorate is added to minimize the **liquid junction potential** and to act as the salt bridge. A standard e.m.f. may be assigned to this half cell in combination with another electrode. Thus at 298 K the standard e.m.f. of the cell

$$\text{saturated calomel electrode in acetic acid} \mid \text{chloranil in acetic acid} \mid \text{Pt}$$

is +0.9095 V, whereas that of the cell

$$\text{saturated calomel electrode in acetic acid} \mid \text{Cl}^- \text{ in acetic acid} \mid \text{AgCl,Ag}$$

is −0.3928 V.

The standard electrode potential for 1 molal calomel electrodes in methanol, aqueous dioxan and aqueous ethylene glycol have been measured. See Ives and Janz (1961) chapter 3.

Capacitance (C; dimensions: εl; units: $F = A^2 s^4 kg^{-1} m^{-2} = A s V^{-1} = C V^{-1}$). Measured by the charge that must be added to a body to raise its potential by one unit. A capacitance of 1 farad requires 1 coulomb of electricity to raise its potential by 1 volt. A conductor charged with a quantity of electricity Q to a potential V has a capacitance $C = Q/V$. The capacitance of a parallel-plate condenser, the area of whose plates is A m^2 and the distance between them d m, is $C = A\varepsilon_0\varepsilon_r/d$.

If $c_1, c_2, \ldots, c_x$ represent the capacitances of a series of condensers and C is their combined capacitance then when connected in parallel

$$C = c_1 + c_2 + \cdots + c_x$$

when in series

$$\frac{1}{C} = \frac{1}{c_1} + \frac{1}{c_2} + \cdots + \frac{1}{c_x}$$

A differential capacitance $\tilde{C}$ may also be defined as

$$\tilde{C} = \left(\frac{\partial Q}{\partial V}\right)_{T,P,\mu}$$

The relation between $\tilde{C}$ and C is

$$\tilde{C} = C\left(1 - \frac{\partial \ln C}{\partial \ln Q}\right)^{-1}$$

Capacitance may be measured from the variation of surface tension of a mercury drop with potential by electrocapillarity (*see* **Electrocapillary phenomena**). A more direct measurement may be made by applying a

32 Carbon monoxide oxidation

galvanostatic step to an electrode and following $\partial V/\partial t$ on an oscilloscope. The differential capacitance is then

$$\tilde{C} = \frac{\partial Q}{\partial V} = \frac{I \partial t}{\partial V} = \frac{I}{\partial V / \partial t}$$

A third method uses a small ac signal superimposed on a dc bias. The frequency of the ac signal chosen is sufficiently high to avoid chemical changes at the electrode, but allows the system to follow the varying potential. A test of the validity of the method is the independence of capacitance with frequency of the ac signal. Adsorption of intermediates gives rise to **pseudocapacitance**, which appears as a large addition to the double-layer capacitance. *See also* **Conductance of aqueous solutions; Electrical double layer**.

Carbon monoxide oxidation. Carbon monoxide may be oxidized electrochemically to carbon dioxide

$$CO + H_2O \rightarrow CO_2 + 2H^+ + 2e \qquad E^\ominus = 1.33 \text{ V}$$

and therefore may be used as fuel for a **fuel cell**. The product, carbon dioxide, will dissolve in an alkaline electrolyte giving carbonates which lower the conductivity of the electrolyte. Carbonate production precludes the use of carbon monoxide, and other organic fuels, in low- and medium-temperature alkaline fuel cells. This problem is turned to advantage in the high-temperature (500°C) molten carbonate cell. The electrolyte is in the form of a paste of molten lithium carbonate, sodium carbonate and potassium carbonate with magnesium oxide. The anodic reaction is

$$CO + CO_3^{2-} \rightarrow 2CO_2 + 2e$$

and at the cathode

$$2CO_2 + O_2 + 4e \rightarrow 2CO_3^{2-}$$

Hydrogen may also be consumed in this cell

$$H_2 + CO_3^{2-} \rightarrow CO_2 + H_2O + 2e$$

and thus reformed natural gas may provide the fuel

$$CH_4 + H_2O \rightarrow 3H_2 + CO$$

Cathode. The electrode where reduction (i.e. gain of electrons, ox + $ne \rightarrow$ red) occurs. *See also* **Anode**.

Cathodic protection. *See* **Corrosion**.

Catholyte. The electrolyte surrounding the **cathode** in an electrochemical cell.

Cation. A positive ion (e.g., H^+, Ca^{2+}). When a current passes through an electrolyte solution, the cations move towards the negative electrode, the **cathode**. *See also* **Anion**.

Cation-selective electrode. *See* **Ion-selective electrode**.

Cell. A series of conducting phases in contact; the electrodes are generally metallic and there are one or more liquid electrolytes. At any phase boundary, where two or more phases of different composition meet, there is a difference of potential. The e.m.f. of the cell is the algebraic sum of all these phase boundary potentials, including any metal contact potential differences that may present. The e.m.f. of a cell is the potential difference between two pieces of metal of identical composition, the ends of the chain of conducting phases.

Several types of cell are recognized:

1) A primary cell is a device in which a spontaneous chemical reaction is used to produce electrical energy. It acts as a source of energy until its materials are exhausted. Examples are the **Daniell cell** and the dry cell.

2) An electrolytic cell is the reverse arrangement, in which electrical energy from an external source brings about the desired chemical reaction (*see* **Electrodeposition of metals; Electrolysis of water**).

3) A secondary cell, or accumulator, is a device that can act as a primary cell until it is exhausted (discharged) and can then be recharged and brought back to its original condition by passing electricity through it from an external source. It acts as an electrolytic cell during the charging process.

Both primary and secondary cells can be connected in series to form a battery, which provides higher voltages than are obtainable from a single cell.

4) A **fuel cell** is a primary cell in which the reagents can be introduced continuously, and the products of the reaction removed continuously so that, in theory, the cell will function for indefinite periods.

5) A **reversible galvanic cell** is specially designed for obtaining thermodynamic data. In any primary cell, the free energy decrease of the reaction taking place can ideally be converted into electrical energy, but in practice 'frictional' losses arise, both in the electrolyte (owing to its resistance) and at the electrodes (**overpotential**). In reversible cells, these losses are avoided by measuring the e.m.f. of the cell under thermodynamically reversible conditions (*see* **Thermodynamics of cells**).

6) A **concentration cell** is a reversible cell in which there is no overall chemical reaction; the e.m.f. results from differences in concentration of one or more of the reactants at the electrode surfaces (*see* **Electrochemical energy conversion; Electrochemical storage**). See Ives and Janz (1961).

Cell constant. *See* **Conductivity**.

34 Chemical potential

Chemical potential (μ; units: J mol^{-1}). Change in the total free energy at constant temperature and pressure when 1 mole of a component is added to an infinite amount of the system.

For the ith component

$$\mu_i = \left(\frac{\partial G}{\partial n_i}\right)_{T,P,n_j}$$

where n_j is the number of moles of the remaining components. For an ion the operation implied by the equation cannot be achieved, since it means adding an ion of one kind only to the solution. The concept is, however, of use in the treatment of the thermodynamics of electrolyte solutions, and at constant temperature the chemical potential of a single ionic species is defined as

$$\mu_i = \mu_i^\ominus + RT \ln a_i \tag{15}$$

where $\mu_i^\ominus$ the standard chemical potential is the chemical potential when $a_i = 1$.

For an electrolyte, the chemical potential can be regarded as the sum of the chemical potentials of the individual ions. For an electrolyte dissociating into ν_+ cations and ν_- anions

$$\mu_2 = \nu_+ \mu_+ + \nu_- \mu_-$$

whence it can be shown that

$$a_2 = a_+^{\nu_+} a_-^{\nu_-} = a_\pm^\nu$$

where a_2 is the activity of the undissociated electrolyte.

To obtain the **electrochemical potential**, the work done in bringing a unit amount of charge from infinity to the region of the given potential has to be added to the chemical potential.

Chemically modified electrode. *See* **Modified electrode.**

Chi potential (χ). *See* **Surface potential.**

Chlorine electrode. *See* **Gas electrode.**

Chlorine production. *See* **Electrolysis of brine.**

Chronoamperometry (chronogalvametry). *See* **Transient methods.**

Chronocoulometry. *See* **Transient methods.**

Chronopotentiometry. *See* **Transient methods.**

Clark cell. *See* **Standard cell.**

Clark oxygen electrode. *See* **Oxygen probe.**

Coated wire ion-selective electrode. A type of **ion-selective electrode** in which an electroactive species is incorporated in a thin polymeric (PVC, epoxy resin) film which is coated directly onto a metallic conductor. These electrodes are small and so the volume of test solution is minimal. Ease of construction of these rugged electrodes makes them suitable for a wide range of applications, such as the determination of NO_x in air and of detergents and amino acids in solution.

Further reading
H. Frieser, Coated wire ion-selective electrodes, in *Ion-Selective Electrodes in Analytical Chemistry*, vol. II, H. Frieser (ed.) Plenum Press, New York, p. 85 (1980)

Colloidal electrolytes. Solutions of highly surface-active materials, such as detergents, soaps and dyes, exhibit unusual physical properties. In dilute aqueous solution, they behave as normal electrolytes, but at a well-defined concentration, the critical micelle concentration (cmc), abrupt changes occur in such physical properties as electrical conductivity, transport number, osmotic pressure and surface tension (*see* Figure 14).

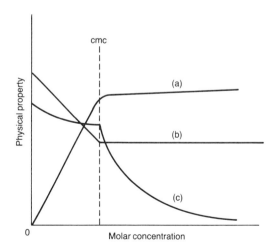

Figure 14. Physical properties of a colloidal electrolyte. (a) Osmotic pressure, (b) surface tension and (c) molar conductivity

36 Complex ions

Above the cmc, the variation of the molar conductivity with concentration shows marked deviations from the Debye–Hückel–Onsager relationship (*see* **Conductance equations**). The osmotic properties of the solution above this concentration are more characteristic of a non-electrolyte. This behaviour is attributed to the formation of micelles by the aggregation of similarly charged ions. The total viscous drag on the surfactant molecules is reduced on aggregation and the number of gegen or counter ions associated with the micelles reduces the number of ions available for carrying the current. This has the effect of lowering the net charge on the micelle, and the retarding influence of the ionic atmosphere is reduced.

For example, the molar conductivity of a solution of the cationic detergent cetylpyridinium bromide, $CH_3(CH_2)_{15}C_5H_4N^+Br^-$, decreases with increasing concentration according to the normal square root relationship to a concentration of about 10^{-3} mol dm^{-3}. Micelle formation, caused by the aggregation of about 68 cetylpyridinium ions, results in a very marked decrease in the molar conductivity. The ions are arranged with the hydrophobic portion of the molecule oriented inwards and the hydrophilic cationic groups oriented towards the aqueous phase. The measured **transport number** of the anion becomes negative at higher concentrations, indicating that bromide ions are being preferentially transported towards the cathode rather than to the anode. It has been calculated that about 53 bromide ions are associated with each micelle, thereby giving the micelle a net charge of $+15$. Under very high field strengths (2×10^7 V m^{-1}), the effects of the ionic atmosphere are removed, leaving the gegen ions virtually free from the micelle, and the molar conductivity rises to very high values.

A low value of the cmc is favoured by an increase in the length of the hydrophobic moiety, a decrease in the temperature and the addition of simple salts. Most evidence is in favour of a spherical shape for the micelles which are essentially liquid droplets of colloidal dimensions with the charged groups situated at the surface. Surfactants at concentrations above the cmc solubilize organic compounds that are insoluble in water (e.g., azobenzene, xylenol orange).

Complex ions. *See* **Conductivity at infinite dilution; Fused salts; Transport number.**

Concentration cell. A **cell** in which there is no overall chemical reaction; the reaction occurring at one electrode (or pair of electrodes) is reversed at the other (or other pair). There may, nevertheless, be a net change of free energy because of a difference in the concentration of one or other of the reactants concerned at the electrodes. The electrical energy arises from the free energy change accompanying the transfer of material from one concentration to the other. The following general types of cell are recognized:

Concentration cells with transport. In these there is direct transfer of ions across liquid junctions.
1) Simple cells containing a **salt bridge** (to minimize the **liquid junction potential**)

$$\ominus \ \text{Ag} \mid \text{AgNO}_3 \mid \text{NH}_4\text{NO}_3 \mid \text{AgNO}_3 \mid \text{Ag} \ \oplus \qquad m_1 > m_2$$
$$(m_2) \qquad\qquad\qquad (m_1)$$

At the right-hand electrode reduction occurs

$$\text{Ag}^+ + e \to \text{Ag}(s)$$

At the left-hand electrode oxidation occurs

$$\text{Ag}(s) \to \text{Ag}^+ + e$$

The net result is the transfer of silver ions from a solution of molality m_1 to one of molality m_2; this results in a free energy change which gives rise to an e.m.f. The electrical neutrality of the solutions is maintained by the diffusion of ions from the salt bridge. Since each electrode is a reversible **silver electrode**, it follows that

$$E = E_{RH} - E_{LH} = \frac{RT}{F} \ln \frac{a_{\text{Ag}^+,1}}{a_{\text{Ag}^+,2}} = \frac{RT}{F} \ln \frac{a_{\pm,1}}{a_{\pm,2}} \tag{16}$$

assuming that

$$a_{\text{NO}_3^-} = a_{\text{Ag}^+}$$

As the reaction proceeds, the concentrations of the two solutions approach each other until both have the same activity when $E_{\text{cell}} = 0$.

Such cells are of use in the determination of the solubility and **solubility product** of sparingly soluble substances.

2) Simple cells in which the liquid junction is not eliminated

$$\ominus \ \text{Ag} \mid \text{AgNO}_3 \mid \text{AgNO}_3 \mid \text{Ag} \ \oplus \qquad m_1 > m_2$$
$$(m_2) \qquad (m_1)$$

For the passage of 1 Faraday of electricity, transfer of material occurs (*see* Table 6). The overall change is the transfer of t_- mole of silver and nitrate ions from m_1 to m_2. From a consideration of the **chemical potential** of the species concerned

$$\Delta G = \sum_i n_i \mu_i$$

$$= t_- RT \ln \frac{(a_{\text{Ag}^+} a_{\text{NO}_3^-})_2}{(a_{\text{Ag}^+} a_{\text{NO}_3^-})_1} = -FE$$

38 Concentration cell

Table 6

Left-hand side	Right-hand side
At electrode $Ag(s) \rightarrow Ag^+ + e$	At electrode $Ag^+ + e \rightarrow Ag(s)$
At junction t_+ mole of Ag^+ lost	At junction t_+ mole of Ag^+ gained
Net gain $1 - t_+ = t_-$ mole Ag^+	Net loss $1 - t_+ = t_-$ mole Ag^+
Gain of t_- mole of NO_3^- by transfer	Loss of t_- mole of NO_3^- by transfer

Thus

$$E = 2t_- \frac{RT}{F} \ln \frac{a_{\pm,1}}{a_{\pm,2}} \tag{17}$$

assuming that the transport numbers of the ions are constant over the concentration range m_1 to m_2. This e.m.f. includes the junction potential. The general equation for the e.m.f. of a cell reversible to cations is

$$E = t_- \frac{\nu}{\nu_+} \frac{RT}{nF} \ln \frac{a_{\pm,1}}{a_{\pm,2}}$$

The ratio of the e.m.f. of the cell in which the liquid junction potential is not eliminated (Eq. (17)) to that for which it is eliminated (Eq. (16)) gives an approximate value for the **transport number** of the ion to which the electrodes are not reversible.

3) Simple cells containing electrodes of the second kind

$$\ominus \ Ag,AgCl \ | \ KCl \ \vdots \ NH_4NO_3 \ \vdots \ KCl \ | \ AgCl,Ag \ \oplus \qquad m_1 > m_2$$
$$(m_1) \qquad\qquad\qquad (m_2)$$

At the right-hand electrode reduction occurs

$$AgCl(s) + e \rightarrow Ag(s) + Cl^-$$

whereas at the left-hand electrode oxidation occurs. The net result is the transfer of chloride ions from m_1 to m_2, for which

$$E = E_{RH} - E_{LH} = \frac{RT}{F} \ln \frac{a_{Cl^-,1}}{a_{Cl^-,2}}$$

$$= \frac{RT}{F} \ln \frac{a_{\pm,1}}{a_{\pm,2}} \tag{18}$$

assuming that

$$a_{K^+} = a_{Cl^-}$$

For the same cell in which the junction potential is not eliminated

$$\ominus \; \text{Ag,AgCl} \; | \; \underset{(m_1)}{\text{KCl}} \; \vdots \; \underset{(m_2)}{\text{KCl}} \; | \; \text{AgCl,Ag} \; \oplus \quad m_1 > m_2$$

$$E = t_+ \frac{RT}{F} \ln \frac{a_{\pm,1}}{a_{\pm,2}} \tag{19}$$

Concentration cells without transport. These are cells in which there are no liquid junctions. There is no direct transfer of ions or electrolyte from one solution to the other; material transport occurs indirectly as a result of chemical reactions. Such cells result when two simple, reversible galvanic cells whose electrodes are reversible with respect to each of the ions constituting the electrolyte are combined in opposition. Consider two cells of the type

$$\text{Pt,H}_2(g) \, | \, \text{HCl(aq)} \, | \, \text{AgCl,Ag}$$

with different concentrations of HCl, m_1 and m_2, where $m_1 > m_2$, each of which has the cell reaction

$$\tfrac{1}{2}\text{H}_2(g) + \text{AgCl}(s) \rightleftharpoons \text{Ag}(s) + \text{H}^+ + \text{Cl}^-$$

If they are now connected in opposition through the silver

$$\ominus \; \text{Pt,H}_2(g) \, | \, \underset{(m_2)}{\text{HCl}} \, | \, \text{AgCl,Ag} \text{—Ag,AgCl} \, | \, \underset{(m_1)}{\text{HCl}} \, | \, \text{H}_2(g)\text{,Pt} \; \oplus$$

The overall reaction, the sum of the two simple cell reactions, consists only of the transfer of HCl from m_1 to m_2 (assuming that the pressure of hydrogen gas at the two terminal electrodes is the same). Thus

$$\Delta G = (\mu_{\text{H}^+,2} - \mu_{\text{H}^+,1}) + (\mu_{\text{Cl}^-,2} - \mu_{\text{Cl}^-,1})$$

$$= RT \ln \frac{a_{\text{H}^+,2} a_{\text{Cl}^-,2}}{a_{\text{H}^+,1} a_{\text{Cl}^-,1}}$$

$$= 2RT \ln \frac{a_{\pm,2}}{a_{\pm,1}} = -FE$$

Hence

$$E = \frac{2RT}{F} \ln \frac{a_{\pm,1}}{a_{\pm,2}} \tag{20}$$

Such cells are used for the determination of activity values and for testing the validity of the **Debye–Hückel activity equation**. If the hydrogen gas pressure differs at the two electrodes, there will be an additional term in Eq. (20) for the transfer of hydrogen gas (*see below*).

40 Concentration cell

Electrode concentration cells.

1) Gaseous electrodes of different pressures in the same solution

$$\text{Pt,H}_2(g) \mid \text{HCl} \mid \text{H}_2(g),\text{Pt}$$
$$(p_1) \qquad\qquad (p_2)$$

The passage of 2 Faradays of electricity is accompanied by the transfer of 1 mole of hydrogen from p_1 to p_2. Thus assuming ideal behaviour

$$E = \frac{RT}{2F} \ln \frac{p_1}{p_2} \qquad (21)$$

For the chlorine gas concentration cell

$$\text{C,Cl}_2(g) \mid \text{PbCl}_2(\text{fused}) \mid \text{Cl}_2(g),\text{C}$$
$$(p_1) \qquad\qquad\qquad (p_2)$$

$$E = \frac{RT}{2F} \ln \frac{p_2}{p_1} \qquad (22)$$

The e.m.f. of this cell is positive when $p_1 < p_2$ (compare with Eq. (21)).

2) Amalgam concentration cells

$$\ominus \; \text{Zn/Hg} \mid \text{ZnSO}_4 \mid \text{Zn/Hg} \; \oplus \qquad x_1 > x_2$$
$$(x_1) \qquad\qquad (x_2)$$

where x_1 and x_2 are the mole fractions of zinc in mercury. At the right-hand electrode reduction occurs

$$\text{Zn}^{2+} + 2e \rightarrow \text{Zn}$$
$$(x_2)$$

and at the left-hand electrode oxidation occurs

$$\text{Zn} \rightarrow \text{Zn}^{2+} + 2e$$
$$(x_1)$$

Since the concentration of zinc ions remains constant, the net change is the transfer of zinc from x_1 to x_2, for which

$$\Delta G = RT \ln \frac{a_2}{a_1}$$

and

$$E = \frac{RT}{2F} \ln \frac{a_1}{a_2}$$

where a_1 and a_2 are the activities of zinc in the two amalgams. For the general case in which the valence of the metal is z and there are y atoms in the molecule

$$E = \frac{RT}{zyF} \ln \frac{a_1}{a_2}$$

$$\approx \frac{RT}{zyF} \ln \frac{c_1}{c_2}$$

when the amalgams are sufficiently dilute. This type of cell is of use in the determination of the activity or concentration of metals in amalgams or alloys.

Concentration cells involving non-aqueous solvents. The e.m.f. of a concentration cell without transference of the type

$$\text{M} \mid \text{M}^+\text{Cl}^- \mid \text{AgCl,Ag—Ag,AgCl} \mid \text{Cl}^-\text{M}^+ \mid \text{M}$$
$$(S_1) \qquad\qquad\qquad\qquad\qquad (S_2)$$

in which the same electrolyte is at the same concentration in the two solvents S_1 and S_2 is given by

$$E = E_2^\ominus - E_1^\ominus - \frac{2RT}{F} \ln \frac{a_{\pm,2}}{a_{\pm,1}}$$

$E_1^\ominus$ and $E_2^\ominus$ are the standard e.m.f. values of the two separate cells. When the two solvents are the same, $E_2^\ominus - E_1^\ominus$ is zero (*see* Eq. (20)). If the solvents are different, $E_2^\ominus - E_1^\ominus$ is not zero but is related to the transfer of MCl from S_2 to S_1; $a_{\pm,2}$ may only be evaluated if E, $E_1^\ominus$, $E_2^\ominus$ and $a_{\pm,1}$ are known.

In concentration cells with transference in which there are two different solvents

$$\text{M} \mid \text{M}^+\text{Cl}^- \mid \text{Cl}^-\text{M}^+ \mid \text{M}$$
$$(S_1) \qquad (S_2)$$

a more complex problem arises since the liquid junction potential depends on the value of the transport number of the chloride ion in each solvent. Further the value of the transport number also depends on the concentration. The experimental attainment of a zero value for the **liquid junction potential** is, therefore, questionable. Thermodynamic data based on cells containing an aqueous/organic solvent liquid junction must be regarded with caution. See Glasstone (1942); Robinson and Stokes (1970).

Concentration overpotential. *See* **Overpotential.**

Conductance (G; dimensions: $\varepsilon l\, t^{-2}$; units: siemens, $\text{S}=\Omega^{-1}=\text{kg}^{-1}\text{m}^{-2}\text{s}^3\text{A}^2$). The reciprocal of electric resistance. The resistance (R) of a conductor is proportional to its length (L) and inversely proportional to its cross-section (A), that is $R = \rho(L/A)$. Thus

42 Conductance at high field strengths

$$G = \frac{1}{R} = \frac{1}{\rho(L/A)} = \frac{\kappa}{L/A}$$

where the **conductivity** (κ) is the reciprocal of the resistivity ρ.

Conductance at high field strengths (Wien effects). The conductance of an electrolyte changes if very high potential gradients are applied. With strong electrolytes there is a moderate increase in conductance, whereas with weak electrolytes there is a very large increase.

1) *First Wien effect*. Under normal potential gradients (up to 100 V m^{-1}) the molar conductivity of a strong electrolyte is independent of the potential gradient, but very large values are observed at very high field intensities (approximately 2×10^6 V m^{-1}). At these high potential gradients the ionic velocities become very large (approximately 0.1 m s^{-1}), and this means that during the relaxation time (approximately 10^{-8} s in 1 mol m^{-3} solution) the central ion moves a distance of about 10^{-9} m (i.e. a distance some ten times the thickness of the ionic atmosphere). The central ion is thus virtually free of its atmosphere and neither the asymmetry nor the electrophoretic effect hinders its motion and therefore the conductance increases. The effect is most pronounced with polyvalent ions because the increase in molar conductivity Λ is proportional to $|z_+||z_-|$ (*see* Figure 15). For dilute solutions, Λ increases and approaches Λ° at high field strengths (*see* Figure 16).

2) *Second Wien effect*. Under intense potential gradients, the increase in molar conductance is more prominent with weak electrolytes due to the

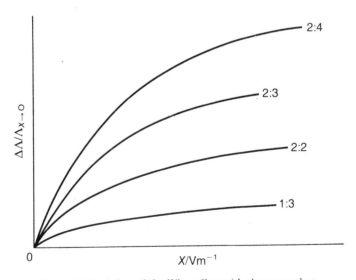

Figure 15. Variation of the Wien effect with charge number

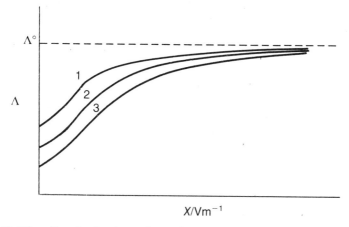

Figure 16. Wien effect showing the tendency of the molar conductance to approach the limiting conductance (dotted line). Curves 1, 2 and 3 represent effects for solutions of three different concentrations $c_3 > c_2 > c_1$

enhancement of the degree of dissociation

$$HA + H_2O \rightleftharpoons H_3O^+ + A^-$$

At high field intensities, the rate of separation of the proton is enhanced but the rate of the reverse process is unchanged. The absence of an ionic atmosphere round the ion reduces the concentration of ions and this, by a mass action effect, favours further ionization of molecules. The additional ions produced give rise to the abnormally high molar conductivities (*see* Figure 17). Onsager has shown, theoretically, that the rate of dissociation of

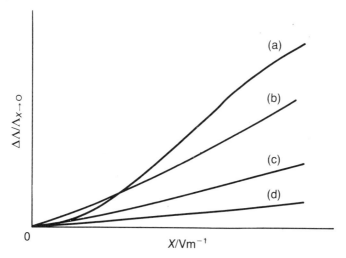

Figure 17. Wien effect with weak acids: (a) acetic acid; (b) tartaric acid; (c) monochloroacetic acid; (d) sulphuric acid

44 Conductance at high frequencies

an **ion pair** is similarly increased whereas the rate of the reverse process is unaffected, thus salts in solvents of low relative permittivity, where ion pairs are present (e.g., potassium iodide in acetone), also show the second Wien effect. See Harned and Owen (1958).

Conductance at high frequencies. Electrolytes show an enhanced conductance when the measurements are made at very high frequencies. The relaxation–time term in the Debye–Onsager equation (*see* **Conductance equations**) decreases and eventually becomes zero as the frequency of measurement increases. The time taken to establish the ionic atmosphere around an ion is about 10^{-6} second in dilute solutions, and if the frequency used for measurement is so high that the ion is oscillating at a similar or greater rate, the atmosphere around the ion will not be fully established and its retarding effect will be reduced.

The effect is most pronounced for ions of low valence and small mobility, and for solvents of low dielectric constant and at low temperatures. See Harned and Owen (1958).

Conductance equations. The interionic attraction theory provides an explanation for the concentration dependence of conductance in very dilute solutions. Debye and Hückel's original treatment was improved by Onsager in 1926, and his equation has been convincingly tested over a wide range of conditions.

If there were no forces between the ions in a dilute solution of a completely dissociated electrolyte (e.g., sodium chloride), thermal motion would keep them randomly and, on average, uniformly distributed throughout the solution. According to Coulomb's law, there are forces of attraction or repulsion between them ($e^2/4\pi\varepsilon_0\varepsilon r^2$); these forces are appreciable over quite large distances. A given positive ion will therefore attract the surrounding negative ions and repel other positive ions, so that at any point close to the central ion there will be a greater likelihood of finding a negative rather than a positive ion. These directive forces are superimposed on the random thermal motions, but a time average distribution can be calculated. Thus each ion has in its close neighbourhood a slight excess of ions of the opposite sign; this is known as the ionic atmosphere. The effects of this atmosphere become less important as the solution becomes more dilute and disappear at infinite dilution.

If the charge on an ion were suddenly annulled, its ionic atmosphere would no longer be attracted and would disperse. This, or the opposite process of building up an atmosphere, takes a finite time depending on the diffusion rates of the ions involved. This is the relaxation time and for dilute solutions is about 10^{-6} second.

The presence of an atmosphere affects the mobility of an ion in two ways. First when an ion moves under an applied field through the liquid, its atmosphere will build up around its new position and will die away around the position it has just left. Since this takes a finite time, the ion will always be moving away from a region containing excess charge of the opposite sign and so will be subjected to a retarding force. This is the relaxation–time effect. The second retarding influence is the electrophoretic effect. The ions in moving through the solution transfer momentum to the solvent molecules, but each ion, on account of its atmosphere, is surrounded by an excess of those moving in the opposite direction. The ions will thus be moving 'up-stream' and will be opposed by a greater frictional force than the stationary solvent would exert. Both retarding effects depend on the density of the ionic atmosphere and increase as the square root of the concentration.

For the molar ionic conductivity (Λ_i/Ω^{-1} cm^2 mol^{-1}) of an ion species in a very dilute solution of a strong electrolyte Onsager derived the expression

$$\Lambda_i = \Lambda_i^\circ - \left(\frac{8\pi N_A e^2}{1000\varepsilon kT}\right)^{1/2} \left(\frac{F^2 z_i}{6\pi\eta N_A} + \frac{|z_+ z_-|}{3\varepsilon kT} \frac{e^2 \Lambda_i^\circ q}{1+\sqrt{q}}\right)\left(\frac{|z_+ z_-|}{2}\right)^{1/2} c^{1/2} \qquad (23)$$

where

$$q = \frac{|z_+ z_-|(\Lambda_+^\circ + \Lambda_-^\circ)}{(|z_+| + |z_-|)(|z_+|\Lambda_-^\circ + |z_-|\Lambda_+^\circ)}$$

For symmetrical electrolytes $|z_+| = |z_-|$, therefore $q = 0.5$.

For a completely dissociated electrolyte, the molar conductivity is the sum of two such expressions. Substituting the values of the universal constants the limiting equation for an electrolyte becomes

$$\Lambda = \Lambda^\circ - \left[\frac{41.25(|z_+| + |z_-|)}{\eta(\varepsilon T)^{1/2}} + \frac{2.801 \times 10^6 |z_+ z_-| q \Lambda^\circ}{(\varepsilon T)^{3/2}(1+\sqrt{q})}\right] I^{1/2} \qquad (24)$$

for a symmetrical electrolyte, Eq. (24) can be written as

$$\Lambda = \Lambda^\circ - (A + B\Lambda^\circ)I^{1/2} = \Lambda^\circ - SI^{1/2} \qquad (25)$$

where S is the Onsager slope and A and B constants with values given by Eq. (24); A gives a measure of the electrophoretic effect and B the relaxation–time effect. The theory is applicable to different solvents, some values for the constants A and B for 1:1 electrolytes are given in Table 7.

Eq. (25) reproduces the limiting slopes of $\Lambda - c^{1/2}$ curves very satisfactorily over a wide range of conditions. For potassium chloride solutions, the calculated slopes agree with the experimental values at 273 K ($S = 47.3$) and at 373 K ($S = 313.4$). The equation also reproduces the large effects of changing the valence type of the electrolyte (*see* Figure 18) and of the variation in the dielectric constant and viscosity of the solvent (*see* **Conductance of non-aqueous solutions**). The equation is accurate at concentrations of up to 2×10^{-3} mol dm^{-3} for 1:1 electrolytes in water. For higher valence type salts and for electrolytes in non-aqueous solvents its range is more restricted.

46 Conductance equations

Table 7
Debye–Hückel coefficients for 1:1 electrolytes at 298 K

Solvent	$A/\Omega^{-1}\,cm^2\,mol^{-1}\,(mol\,dm^{-3})^{-1/2}$	$B/(mol\,dm^{-3})^{-1/2}$
Water	60.20	0.229
Methanol	156.1	0.923
Ethanol	89.7	1.83
Acetonitrile	22.9	0.716
Acetone	32.8	1.63
Nitromethane	125.1	0.708
Nitrobenzene	44.2	0.776

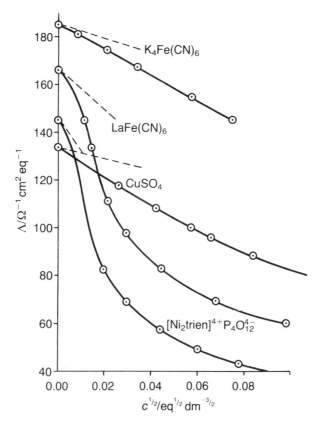

Figure 18. Equivalent conductivities of some multivalent salts in water at 298 K. (Dashed lines, Onsager slopes)

The success of the equation suggests that the model of ion–ion interactions is substantially correct. Further tests of the theory can be obtained from studies of the conductance of solutions when the effect of the ionic atmosphere is eliminated: for example, **conductance at high frequencies** and **conductance at high field strengths**.

The Onsager equation also applies to weak electrolytes, provided that the concentration (c) is replaced by αc where α is the degree of ionization, and to strong electrolytes in which the **ion pair** is incompletely dissociated, in which case α is the degree of dissociation of the ion pair.

Extensions to higher concentrations. The extension of the theory to higher concentrations has been undertaken by Falkenhagen and his coworkers, by Pitts and by Onsager and Fuoss. The theory is very difficult and the final equations inevitably involve assumptions and mathematical approximations that have been the subject of much controversy. The equations proposed can satisfy the best experimental results for 1 : 1 electrolytes up to 0.1 mol dm^{-3}; for un-symmetrical valence types, the mathematical problems are intractable, and for any salt above 0.1 mol dm^{-3} the model on which the theory is based becomes unreliable.

The main requirement in extending the theory is the introduction of a parameter a, the mean distance of closest approach or mean ionic diameter, to allow for the fact that ions have finite size and are not the point charges assumed in deriving the limiting laws. The value of a is chosen to give the best fit with the experimental data, the only requirement being that it should be a plausible value (approximately 10^{-10} m) for the effective diameter of the dissolved ion. The different theories show variations in their precise definition of a, and require different values to satisfy the same data.

For practical purposes, the most useful equation is that of Robinson and Stokes, who have shown that the Falkenhagen equation can be approximated to

$$\Lambda = \Lambda^\circ - \frac{Sc^{1/2}}{1 + \kappa a} \tag{26}$$

which is simply the limiting equation with the Onsager term divided by $(1 + \kappa a)$, where κ, the reciprocal thickness of the ionic atmosphere, is given by

$$\kappa = \left(\frac{8\pi N_A e^2}{1000\varepsilon kT}\right)^{1/2} I^{1/2} \tag{27}$$

For 1 : 1 electrolytes in water at 298 K, κ (in units of m^{-1}) is given by

$$\kappa = 0.3291 \times 10^{10} \, (I/\text{mol dm}^{-3})^{1/2}$$

48 Conductance equations

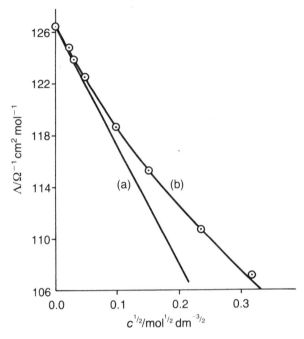

Figure 19. Molar conductivity of sodium chloride at 298 K as a function of concentration according to (a) Onsager's equation and (b) Robinson and Stokes's equation ($a = 4 \times 10^{-10}$ m).

Data for sodium chloride plotted according to Eq. (26) are shown in Figure 19.

Shedlovsky observed that for strong 1:1 electrolytes the value of Λ° calculated from individual values of Λ using Eq. (25) which can be rearranged to

$$\Lambda^\circ = \frac{\Lambda + A\sqrt{c}}{1 + B\sqrt{c}} \qquad (28)$$

is not constant as required by the limiting law but varies almost linearly with concentrations of up to 0.1 mol dm^{-3}. To extend the limiting equation (Eq. (26)) to higher concentrations he proposed the equation

$$\Lambda = \Lambda^\circ - (A + B\Lambda^\circ)c^{1/2} + Cc + Dc^{3/2} \qquad (29)$$

where C and D are empirical constants. In many cases D is negligible and Eq. (29) becomes

$$\Lambda^\circ = \frac{\Lambda + Ac^{1/2}}{1 - Bc^{1/2}} - Cc \qquad (30)$$

See Covington and Dickinson (1973); Robinson and Stokes (1970).

Further reading

H. P. Bennetto, Electrolyte solutions, *Ann. Rep. Chem. Soc.*, **70**, 223 (1973)

D. F. Evans and M. A. Matesich, The measurement and interpretation of electrolytic conductance, in *Techniques of Electrochemistry*, vol. 2, E. Yeager and A. J. Salkind (ed.), Wiley–Interscience, New York (1973)

J.-C. Justice, Conductance of electrolyte solutions, in *Comprehensive Treatise of Electrochemistry*, vol. 5, J. O'M. Bockris, *et al.* (ed.), p. 223 (1982)

Conductance minima. Occasionally the curve of molar conductivity against the square root of the concentration passes through a minimum value. This behaviour is common for salts in pyridine, liquid sulphur dioxide and the aliphatic amines (all solvents with dielectric constants of about 10). It also occurs in aqueous solutions of lanthanum ferricyanide and of the **colloidal electrolytes**.

Since there is no reason for thinking that ionic conductivities could increase with concentration, the explanation must lie in a change in the number of ions present. The dissociation constant of a partially dissociated binary electrolyte

$$MA \rightleftharpoons M^{z+} + A^{z-}$$

is given by

$$K = \frac{\gamma_+ \gamma_- \alpha^2 c}{1 - \alpha} \qquad (31)$$

If, over a given concentration range, the activity coefficients decrease more rapidly than the increase in the concentration, the value of α would have to increase to maintain equilibrium, the rapid fall in the activity coefficients of the ions being offset by an increase in their number. The increase in α would then tend to cause an increase in the conductivity.

The Debye–Hückel equation for the mean ionic activity coefficient is

$$-\log \gamma_\pm = \frac{A(\alpha c)^{1/2}}{1 + B\mathring{a}(\alpha c)^{1/2}} \qquad (32)$$

The coefficient $\gamma_\pm$ can therefore be evaluated, and for $\mathring{a} = 4 \times 10^{-10}$ m at 298 K, Eq. (31) and Eq. (32) predict a minimum in the degree of dissociation at $\alpha c = 5 \times 10^{-3}$ mol dm^{-3} for a solvent of dielectric constant of 12 and at $\alpha c = 1 \times 10^{-3}$ mol dm^{-3} for a solvent of dielectric constant of 7. This therefore provides an explanation for the findings in non-aqueous solvents.

In water, no dissociation minimum would be possible on this basis for a 1:1 electrolyte, but in lanthanum ferricyanide, with a valency product of 9, the interionic forces will be comparable with those of a 1:1 salt in a medium of dielectric constant 9, and the same expanation is applicable (*see* Figure 20).

Conductance minima

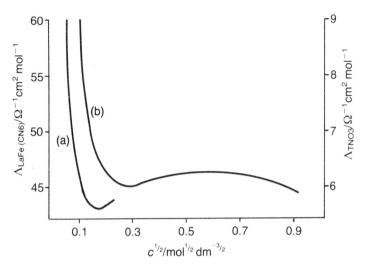

Figure 20. Variation of the molar conductivity with concentration for (a) tetra-iso-amylammonium nitrate (TNO$_3$) in water/dioxane (water 14.95 percent, $\varepsilon = 8.5$) and (b) lanthanum ferricyanide in water

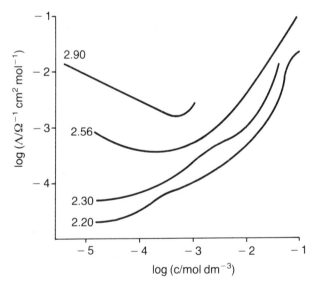

Figure 21. Conductance of tetra-iso-amylammonium nitrate in dioxane/water mixtures. The dielectric constant of the solvent is shown against curves

In solvents of very low dielectric constant (relative permittivity) the increase in molar conductivity with concentration (*see* Figure 21) has been attributed by Fuoss and Kraus to the formation of triple ions of the type $(+-+)$ or $(-+-)$ followed by further association to quadrupoles or higher aggregates. Assuming the equilibria

$$MA \rightleftharpoons M^+ + A^-$$
$$MA_2^- \rightleftharpoons MA + A^-$$
$$M_2A^+ \rightleftharpoons MA + M^+$$

it can be shown that when the degree of dissociation of the last species is small

$$\Lambda = k_1 c^{-1/2} + k_2 c^{1/2}$$

where k_1 and k_2 are constants. This equation predicts that the molar conductivity will pass through a minimum at a certain concentration and thereafter increases with increasing concentration. For tetra-iso-amylammonium nitrate, triple ions cannot theoretically be stable in solvents of dielectric constant >23.2; experimental data are in accordance with this prediction. Further evidence in support of the presence of complex aggregates is provided from relative molecular mass determinations by the cryoscopic method; at 4×10^{-3} mol dm^{-3} the relative molecular mass is twice the formula weight and increases with increasing concentration. See Davies (1962); Dole (1935); Harned and Owen (1958).

Conductance of aqueous solutions. The **conductivity** (κ) of an electrolyte increases with increasing concentration because of an increase in the total number of ions present. The increase in ionic concentration may be the result of an increase in the electrolyte concentration or to changes in the extent of ionization (α) of the electrolyte.

The variation of the molar conductivity (Λ) with concentration reveals that there are two extreme groups of electrolytes: strong and weak (*see* Figure 22).

1) Weak electrolytes (e.g., acetic acid, aqueous ammonia) are poor conductors because the ionization is small; the equilibrium

$$H_2O + HA \rightleftharpoons H_3O^+ + A^-$$

lies to the left. Ionization increases as the concentration decreases and the molar conductivity increases rapidly to a very high value at zero concentration (for acetic acid, $\Lambda^\circ = 390.4 \times 10^{-4}\ \Omega^{-1}\ m^2\ mol^{-1}$). Accurate values of Λ° for weak electrolytes cannot be obtained by extrapolation but must be obtained from the sum of the individual **molar ionic conductivity** values (*see* **Conductivity at infinite dilution**).

2) Strong electrolytes (e.g., salts of alkali metals, strong mineral acids and alkalis, quaternary ammonium salts $NR_4^+\ A^-$) exhibit a much greater con-

Conductance of aqueous solutions

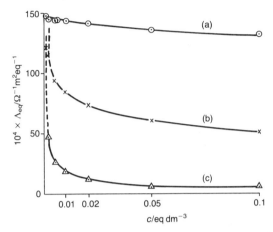

Figure 22. Variation of equivalent conductance with concentration. (a) Potassium chloride, (b) nickel sulphate and (c) acetic acid

ductivity than do weak electrolytes since they are completely ionized, even in the solid state. On solution in water the ions become hydrated and the hydrated ions disperse in the medium. The increase in molar conductivity with decreasing concentration is explained by variations in the interionic forces of attraction (*see* **Conductance equations**). At low concentrations, the plot of Λ against the square root of the concentration is linear; the slope of the line depends on the valence type of the electrolyte. For 1 : 1 electrolytes (*see* Figure 23), the slope is in good agreement with the values calculated from the Onsager's equation (*see* Eq. (24)). Strong electrolytes in water are not necessarily strong electrolytes in other solvents.

It is impossible to draw a sharp division between the two categories of electrolytes for many solutes exhibit intermediate behaviour: for example, the divalent metal sulphates (nickel sulphate, *see* Figure 22) in water (*see* **Ion pair**) and other electrolytes in solvents of low dielectric constant (*see* **Conductance of non-aqueous solutions**).

Conductance of mixtures. In a mixture of electrolytes, all the ions present contribute to the ionic atmosphere of each ion, and it is the total concentration that governs the conductivity decrease.

The conductance of mixtures always tends to be slightly less than the additive value given by the simple mixture rule. This is because the appearance in the ionic atmosphere of an ion of higher conductivity reduces the relaxation time effect, and vice versa. The effect is greater for the ion of higher conductivity, so the net result is a small decrease in conductance.

If the electrolytes are not completely dissociated, mixing will alter the

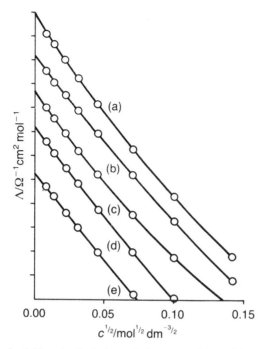

Figure 23. Molar conductivities of uniunivalent salts in water at 291 K. (a) Potassium chloride, (b) potassium fluoride, (c) sodium chloride, (d) sodium nitrate and (e) potassium iodate

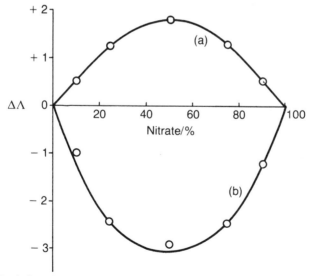

Figure 24. Deviations from additivity of molar conductivity in Na,K–Cl,NO$_3$ mixtures. (a) Sodium chloride–potassium nitrate and (b) potassium chloride–sodium nitrate

54 Conductance of fused salts

number of ions free to conduct and there will be further effects which can be calculated if the dissociation constants are known. This is illustrated in Figure 24 for various mixtures of the reciprocal salt pair sodium chloride–potassium nitrate at a total concentration of 1 mol dm^{-3}. Ion pairing is appreciable in potassium nitrate solutions and much less in sodium nitrate, but absent in the chlorides. The main effect of adding sodium nitrate to potassium chloride is therefore to produce non-conducting potassium nitrate ion pairs, whereas diluting potassium nitrate with sodium chloride leads to some dissociation of the nitrate. The continuous curves are calculated from the known dissociation constants. Much larger effects occur with weaker electrolytes. See Davies (1962); Harned and Owen (1958); Robinson and Stokes (1965).

Further reading
M. J. Wootten, The conductance of electrolyte solutions, in *Electrochemistry*, vol. 3, Specialist Periodical Reports, Chemical Society, London, p. 20 (1973)

Conductance of fused salts. *See* **Fused salts.**

Conductance of non-aqueous solutions. The properties of a solvent which influence the conductance of a salt are its viscosity, its dielectric constant and its interaction with the ionic species present. In protic solvents, hydrogen bonding, especially for small anions, contributes appreciably to the solvent–ion interactions. In aprotic solvents, only ion–dipole and mutual polarizability are of importance in determining solvent–ion interactions.

In protic solvents, Onsager's equation (*see* **Conductance equations**) is valid at very low concentrations for a range of solvents. The conductance equation

$$\Lambda = \Lambda^\circ + Sc^{1/2} + Ec \ln c + J_1 c - J_2 c^{3/2} \tag{33}$$

where E, J_1 and J_2 are complex functions involving such factors as the closest distance of approach of two ions. Eq. (33) is capable of representing the behaviour of completely dissociated electrolytes. Deviations of the experimental data from predicted values are attributed to the formation of neutral ion pairs which do not contribute to the conductance. In sulpholane, the halides of lithium show an increasing deviation from the iodide (slightly dissociated) to the chloride (a typical weak electrolyte). Ion pair association constants can be obtained from conductance measurements. For the symmetrical electrolytes

$$M^{z+} + A^{z-} \rightleftharpoons MA$$

$$K_A = \frac{1 - \alpha}{\alpha^2 c \gamma_\pm^2}$$

where $\gamma_\pm$ is the mean ionic activity coefficient of the free ions at the concentration αc. For the associated electrolyte, Eq. (33) becomes

$$\Lambda = \Lambda^\circ + S(\alpha c)^{1/2} + E(\alpha c) \ln(\alpha c) + J_1 \alpha c - J_2(\alpha c)^{3/2} - K_A \Lambda^\circ \gamma_\pm^2 (\alpha c)$$

In methanol, alkali metal salts (except nitrates) are essentially dissociated. These solutions show a general behaviour which can be accounted for by an electrostatic model in which the ions are considered to be solvated. The values of K_A for quaternary alkylammonium halides in alcohols are larger for iodides, suggesting solvation of the anions through hydrogen bonding. In general, quaternary alkylammonium iodides are more associated in alcohols than in dipolar aprotic (i.e. non-hydrogen-bonded) solvents of similar dielectric constant.

Dipolar aprotic solvents behave as differentiating solvents with the polarizability of the anions largely determining their interaction with the solvent molecule. For the quaternary ammonium compounds in acetone, the value of K_A decreases with the size of the cation and for anions in $(n\text{-Bu})_4\text{NX}$, K_A decreases in the order bromide > iodide ≈ nitrate > chlorate > picrate.

In solvents with dielectric constant of less than 12, electrostatic ionic interactions are very large and often triple ion formation is encountered. The conductance has been observed to pass through a minimum value (*see* **Conductance minima**). See Covington and Dickenson (1973).

Conductiometric titrations. *See* **Electrometric titrations.**

Conductivity (κ; dimensions: $\varepsilon\, t^{-1}$; units: $\Omega^{-1}\, \text{m}^{-1}$). The current flowing across unit area per unit potential gradient; formerly called the specific conductance. It is the reciprocal of the resistivity (ρ).

$$\kappa = \frac{1}{\rho} = \frac{L/A}{R} = JG \qquad (34)$$

In conductance cells, the term L/A is known as the cell constant (J) and must be determined using a solution of known conductivity. The conductivity varies from high values for good conductors to low values for insulating materials (*see* Table 8).

Electrolyte solutions, like other conductors, obey Ohm's Law. Consider a column of electrolyte of cross-section $1\, \text{m}^2$ to which a potential gradient of $1\, \text{V}\, \text{m}^{-1}$ is applied (*see* Figure 25). The current passing will depend on the number and speed of the ions, and on the charges carried by them. All the ions present contribute, and the total current will be the sum effect of the cations migrating towards the cathode and the complementary effect of negatively charged ions moving in the opposite direction. If a cation carrying $|z_+|$ charges is present at a concentration of $c_+\, \text{mol}\, \text{m}^{-3}$, there will be c_+ mole in a 1 metre length and if their velocity under unit potential gradient is $u_+\, \text{m}\, \text{s}^{-1}$, the

Conductivity

Table 8
Conductivity of various materials at 298 K

Material	Conductivity/$\Omega^{-1} m^{-1}$
Silver	6.33×10^7
Copper	5.8×10^7
Fused sodium chloride	3.3×10^2
Potassium chloride (aq, 0.1 mol dm^{-3})	1.33
Acetic acid (aq, 0.1 mol dm^{-3})	5.2×10^{-2}
Water	4.0×10^{-6}
Sulphur	2.5×10^{-14}

number of moles crossing a reference plane normal to the direction of the current in 1 second is $c_+ u_+$. The quantity of electricity associated with 1 mole is $|z_+|F$; thus the total contribution to the current is $|z_+|c_+ u_+ F$. The total conductivity of the solution is the sum of all such contributions

$$\kappa = |z_+|c_+ u_+ F + |z_-|c_- u_- F \tag{35}$$

The **molar ionic conductivity** is defined by

$$\Lambda_i = |z_i| u_i F$$

therefore

$$\kappa = c_i(\Lambda_+ + \Lambda_-) \tag{36}$$

or, in general,

$$\kappa = \sum_i c_i \Lambda_i \tag{37}$$

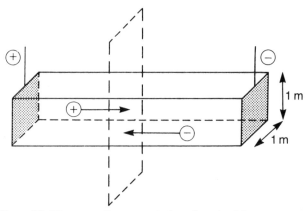

Figure 25. Diagrammatic representation of conductivity

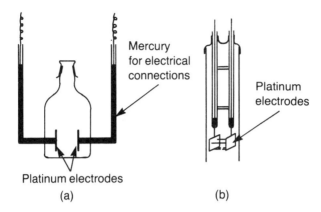

Figure 26. Conductance cells: (a) bottle type; (b) dip type

The conductivity depends on the ionic concentrations and tends to zero as the solution becomes more dilute.

A more convenient unit is the **molar conductivity** (Λ). This is the conductivity of a solution at a concentration of 1 mol m^{-3}; $\Lambda = \kappa/c$.

Measurement of conductivity. The conductance (G) of a solution can be measured using an ac bridge which avoids electrode reactions and polarization of the electrodes; a balance is sought, manually or automatically, by adjusting a variable resistance and capacitance in one arm of the bridge. The conductance cell forms the other arm of the bridge. The resistance of the cell arises from the electrolyte solution and the capacitance from the Helmholtz double layer adjacent to the electrode surfaces.

Conductance cells of the type shown in Figure 26, with platinum disc electrodes of cross-sectional area A separated by a distance L, have to be calibrated (cell constant, J) using an electrolyte solution of known molar conductivity

$$J = \frac{\kappa_{\text{solute}} + \kappa_{\text{solvent}}}{G_{\text{measured}}} \tag{38}$$

A solution of potassium chloride is the accepted reference solution. At 298 K, the molar conductivity of a solution of concentration c mol m^{-3} may be calculated from

$$\Lambda_{\text{KCl}}/\Omega^{-1}\,\text{m}^2\,\text{mol}^{-1} = 149.93 \times 10^{-4} - 2.992 \times 10^{-4}\,c^{1/2}$$
$$+ 58.74 \times 10^{-7}\,c\,\log c + 22.18 \times 10^{-7}\,c \tag{39}$$

Hence a calculated value of $\kappa = \Lambda c$ is obtained. The potassium chloride must be carefully purified by repeated recrystallization from conductivity water and fusing in a platinum crucible. The conductivity of the water used in the

58 Conductivity at infinite dilution

preparation of the solution contributes to the total conductivity and must be allowed for (Eq. (38)); this is of particular importance in very dilute solutions. Carbon dioxide must be rigorously excluded from water and electrolyte solutions. The conductance of electrolyte solutions increases by more than 2 percent per degree, thus close temperature control of the cell is essential.

Conductivity at infinite dilution (Λ°; units: Ω^{-1} m^2 mol^{-1}). The maximum theoretical value of the molar conductivity of an electrolyte, which is attained when the solution is diluted indefinitely with an inert solvent.

Determination of Λ°

1) For most completely dissociated (strong) electrolytes, the linear plot of the molar conductivity against the square root of the concentration (from 10^{-3} mol dm^{-3} downwards) may be extrapolated to zero concentration. The slope of the Λ/c plot should be checked against the theoretical Onsager slope (*see* **Conductance equations**).

2) Values of the molar conductivity measured over a wider range of concentration may be used in Shedlovsky's equation (Eq. (28)), and the linear plot of $(\Lambda + Ac^{1/2})/(1 - Bc^{1/2})$ against the concentration extrapolated to zero concentration.

3) For incompletely dissociated (weak) electrolytes, the above extrapolation methods are not applicable. Use is made of Kohlrausch's law of independent ionic conductivities, that is the addition of the individual **molar ionic conductivity** values gives the limiting molar conductivity of the electrolyte.

$$\Lambda^\circ = \nu_+ \Lambda^\circ_+ + \nu_- \Lambda^\circ_-$$

Alternatively, the values of the limiting molar conductivities of the appropriate strong electrolytes may be used: for example, for the weak acid HA

$$\Lambda^\circ_{HA} = \Lambda^\circ_{KA} + \Lambda^\circ_{HCl} - \Lambda^\circ_{KCl}$$

The molar conductivity at infinite dilution of an electrolyte is characteristic of the number of ions per mole of electrolyte (*see* Table 9). Such data can be used to support the structure of complex ions. Thus if $K_3Fe(CN)_6$ existed in solution as simple ions, the calculated value of $\Lambda^\circ = 3\Lambda^\circ_{K+} + \Lambda^\circ_{Fe2+} + 6\Lambda^\circ_{CN-} = 819 \times 10^{-4}$ Ω^{-1} m^2 mol^{-1}, a value which is not in agreement with the experimental value of 518×10^{-4} Ω^{-1} m^2 mol^{-1}.

Contact adsorption. *See* **Electrical double layer.**

Contact potential. The equilibrium potential difference established when two dissimilar metals M_1 and M_2 are in contact at a given temperature. It is related to the work function (ϕ), that is the work required to extract electrons from the metals, by

Table 9
Molar conductivity values at infinite dilution for electrolytes of different valence type

Valence type	$10^4 \Lambda°/\Omega^{-1} m^2 mol^{-1}$	Examples
1:1	110–150	KBr 151.7 LiCl 115
1:2, 2:1	250–300	$BaCl_2$ 280
1:3, 3:1	400–520	$Fe(NO_3)_3$ 418 K_3PO_4 501 $K_3Fe(CN)_6$ 518
1:4, 4:1	>700	$K_4Fe(CN)_6$ 738

$$V_{M1,M2} = \phi_{M1} - \phi_{M2}$$

$\phi_{Pt} = 4.52$ eV $> \phi_{Cd} = 4.00$ eV, so that less energy is required to remove an electron from cadmium than from platinum, and more energy is released when the electron drops into the platinum lattice than when it drops into the cadmium lattice. Transfer of an electron from cadmium to platinum at the same potential results in the spontaneous release of 0.52 eV of energy, that is the contact potential of cadmium relative to platinum is -0.52 eV. This net transfer of electrons makes the platinum negative with respect to cadmium. Thus, at equilibrium, when electron flow is the same in both directions, the equilibrium potential is the contact potential. If a third metal is introduced between the other two, the potential difference between the two end metals will be the same as if they were directly in contact.

Values of contact potentials listed for different metals are with reference to a standard metal, usually platinum. Contact potentials are extremely sensitive to surface contamination (e.g., oxide formation, adsorption of polar vapours); for this reason they are of use in adsorption phenomena.

Copper coulometer. *See* **Coulometer.**

Copper electrometallurgy. Copper can form two series of salts, and thermodynamic data give the following equilibrium potentials

$$E_{Cu2+,Cu} = +0.337 + \frac{RT}{2F} \ln a_{Cu2+}$$

60 Copper electrometallurgy

and

$$E_{Cu^+,Cu} = +0.521 + \frac{RT}{F} \ln a_{Cu^+}$$

When a solution containing copper ions is in contact with a copper electrode, the system will move spontaneously towards a position of equilibrium through a transfer of electrons between the species Cu^{2+}, Cu^+ and Cu^0. A potential E is thus established which satisfies both equations, and

$$0.337 + \frac{RT}{2F} \ln a_{Cu^{2+}} = 0.521 + \frac{RT}{2F} \ln a_{Cu^+}^2$$

or

$$0.184 = \frac{RT}{2F} \ln \frac{a_{Cu^{2+}}}{a_{Cu^+}^2} = \frac{RT}{2F} \ln \frac{1}{K} \tag{40}$$

where K is the equilibrium constant for the reaction $Cu^{2+} + Cu \rightleftharpoons 2Cu^+$, which comes to 6.4×10^{-7}. If the solution is of unit activity with respect to copper (II) ion, Eq. (40) yields a value for the activity of copper (I) ion of 8×10^{-4} mol dm^{-3}. Copper dissolving at an anode under equilibrium conditions would give the two ions in the ratio required by the equilibrium constant.

Electrolysis is used for extracting copper from poor ores, and for reclaiming scrap copper. The copper is brought into solution as copper (II) sulphate, and this is electrolysed with a cathode of sheet copper and a lead alloy anode. Oxygen is liberated at the anode and sulphuric acid accumulates in the electrolyte. When the electrolyte is exhausted, the sulphuric acid is used to bring more copper into solution.

Owing to the demand for very pure copper, electrolytic refining is practised on a large scale. The cathodes are thin sheets of copper and the anodes blocks of the impure metal, and the electrolyte consists of copper (II) sulphate and sulphuric acid; the presence of the latter improves the conductance and has other advantages. The current density is kept below the value at which hydrogen might be simultaneously discharged, but there is an appreciable **overpotential** at both electrodes. One result of this is that copper (I) ions are formed at the anode at concentrations greater than that corresponding to equilibrium in the bulk of the solution. As the ions diffuse away from the anode, therefore, the process $2Cu^+ \rightarrow Cu^{2+} + Cu$ will tend to occur, leading to the deposition of powdery copper in the bath. The acid electrolyte enables this to be brought back into the solution by the reaction

$$Cu + H_2SO_4 + \tfrac{1}{2}O_2 \rightarrow CuSO_4 + H_2O$$

Of the impurities in the anode, silver and other noble metals pass into the anode sludge; selenium, tellurium, arsenic, antimony and bismuth are also precipitated as insoluble compounds. Other impurities, especially nickel and

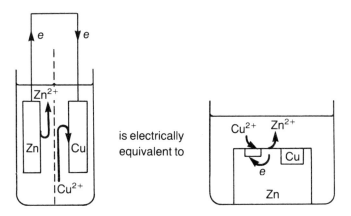

Figure 27. Corrosion of zinc by a local cell which is equivalent to a Daniell cell

iron, pass into solution and gradually accumulate in the electrolyte, which has eventually to be renewed. A bipolar system is sometimes used (*see* **Bipolar electrode**), in which a number of plates of impure copper, not connected to the electricity supply, are spaced out between cathode and anode. On each bipolar electrode impure copper dissolves from one face while pure copper is deposited on the other, so that they eventually consist of pure metal. The energy requirement is much less for this than for the simple arrangement, but there is some leakage of current around the plates. Strict control is also necessary, as an insufficient passage of current would leave layers of impure metal, while any excess of current would be entirely wasted.

Corrosion. The oxidation and dissolution of a metal, usually by the formation of a short-circuited electrochemical cell.

Consider a **Daniell cell** with the anode and cathode directly connected. Zinc corrodes to zinc ions, while copper is plated at the cathode. It is possible to visualize this cell consisting of a bar of zinc with small inclusions of copper (*see* Figure 27). When the bar is immersed in an electrolyte containing copper ions, zinc ions will pass into solution. The electrons released will move through the metal to the more positive regions where copper ions will be discharged by them, and so the corrosion of the zinc will continue. Some practical examples of corrosion occur in this way, when two different metals are in electrical contact with a surface film of moisture to provide the electrolyte. The presence of a second metal, however, is not necessary; local regions of relatively positive or negative potential may result from differences in the activities of surface atoms. The potential of a cold-worked metal is usually slightly more negative than an annealed metal and, in the same piece of metal, grain boundaries or two different crystal faces can provide the two poles of a cell.

62 Corrosion

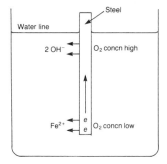

Figure 28. Corrosion in region remote from water-line

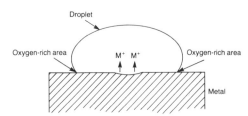

Figure 29. Differential aeration causing corrosion in water droplet

Heterogeneities in the metal surface are not necesary, however, for corrosion to occur. Electrolytic action requires an anode in electrical contact with a cathode and an electrolyte to transport the current through the moisture film or other liquid phase. The anode reaction in corrosion is the dissolution of metal. The cathode reaction need not be the discharge of a metal ion, as in the first example considered above. In aqueous solutions, the possible reactions include the reduction of oxygen, evolution of hydrogen and the reduction of impurities (*see* **Hydrogen electrode reactions; Oxygen electrode reactions**). So long as a pure metal surface is at a potential more positive than its standard electrode potential and more negative than the equilibrium potential of any one of these possible cathodic processes, both cathodic and anodic reactions can occur spontaneously, and the metal will corrode at random points over the surface.

In addition to these effects, corrosion can be caused because of local variations in the solution concentrations, by a type of **concentration cell** effect. A common example of this is illustrated in Figure 28 for a steel structure immersed in sea water. The concentration of oxygen decreases with increasing depth, and so will the equilibrium potential for oxygen reduction. As a consequence, the metal near the surface will act as a cathode, and metal dissolution will occur in the parts further from the water line. The same effect can occur in the case of buried pipes if the availability of oxygen varies from place to place. Another illustration of the effect is given in Figure 29 where an isolated droplet of moisture is shown on the surface of a metal that can be corroded. Oxygen is more readily available at the periphery of the drop and will be reduced here, whereas metal ions will dissolve at the centre of the drop and eventually precipitate, in most cases, as the hydroxide.

Rates of corrosion reactions. Reference to tables of electrochemical potentials or a **Pourbaix diagram** shows the thermodynamic feasibility of corrosion reactions, but the kinetics of the process may be described by reference to

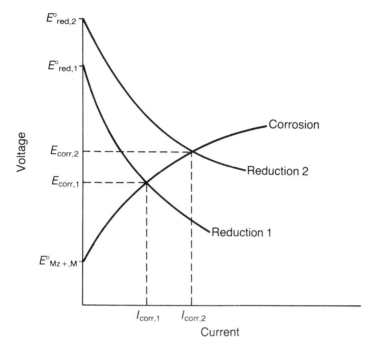

Figure 30. Evans' diagram of the corrosion of a metal showing the effect of different reduction reactions. $E°$ is the potential of the half cell when no current flows

Evans' diagrams (*see* Figure 30). On an Evans' diagram, current–potential plots of the corrosion and reduction reactions are superimposed. At the crossing point, both the potential of the corroding metal and the corrosion current are found.

Prevention of corrosion. The possibilities include control of the environment (perhaps by humidity control, de-aeration or addition of inhibitors which retard the cathodic process) and control of the surface (by surface coatings, electroplating or producing surface layers of high resistance, as in the phosphate treatment of steels). There are also three more purely electrochemical methods.

1) The metal to be protected is connected to a more active metal, that is one with a more negative $E^{\ominus}$ value. The anodic reaction will then occur preferentially at the surface of this more active metal, which will gradually go into solution, thus providing protection. This is an example of sacrificial protection.

2) A potential from an outside source is imposed on the metal to be

64 Corrosion

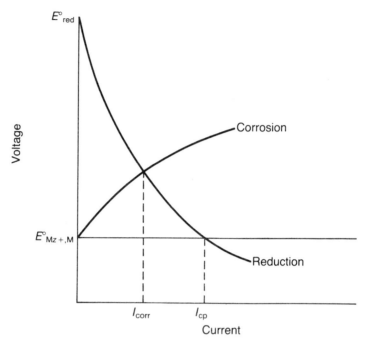

Figure 31. Cathodic protection of a corroding metal. I_{cp} = cathodic protection current

protected, to ensure that the potential at its surface is never more positive than its $E°$ value (i.e. the potential when no current flows). Under these conditions, the metal will not corrode. The energy requirements for this method can be calculated for a particular case by reference to an Evans' diagram (*see* Figure 31). The electrons used in the cathodic process are now being supplied not by the corroding metal but by the outside source. The value at which the required conditions will be met are found by continuing the reduction curve of Figure 31 down to the point at which the potential has become equal to $E°_{Zn2+,Zn}$. The corresponding current density is the minimum current that must pass (between the protected metal and an inert, secondary electrode) to supply the metal with electrons sufficiently quickly to nullify its own tendency to dissolve. These maintain the cathodic reaction, while the potential of the metal is held at its $E°$ value.

3) For metals that passivate (*see* **Passivity**), an anodic potential may be used to form a protective oxide film. The anodic protection current is usually smaller than that required for cathodic protection, and the possibility of undesirable **hydrogen embrittlement** is avoided. See Evans (1960); Fried (1973).

Coulomb (C; dimension, $\varepsilon^{1/2} m^{1/2} l^{3/2} t^{-1}$; units C = A s). The SI unit of charge, defined as the quantity of electricity transported in 1 second by a current of 1 ampere. *See also* **Faraday's laws**.

Coulometer. The quantity of electricity passed through a cell may be determined by measuring the current as a function of time and determining the area under the current–time curve. Alternatively a chemical coulometer, consisting of an electrolytic cell in series with the experimental cell, may be used to measure the same amount of electricity which passes through both cells. The chemical reaction at the anode or cathode must occur with 100 percent efficiency and must be capable of easy and accurate estimation.

In the silver coulometer, which consists of an inert platinum cathode and silver anode dipping into a solution of silver nitrate, the weight of metallic silver deposited on the cathode or the decrease in weight of the anode is determined. The accuracy of this method is limited by the care of manipulation and accuracy of weighing ($1.118\,00 \times 10^{-3}$ g of silver is deposited by 1 C); this is probably the most accurate coulometer.

The iodine coulometer, which makes use of the reaction

$$\tfrac{1}{2}I_2(s) + e \rightleftharpoons I^-$$

consists of platinum/iridium electrodes immersed in potassium iodide solution (anode) and a standard solution of iodine in potassium iodide (cathode); the two compartments are separated by a narrow tube containing potassium iodide solution to prevent diffusion. During electrolysis iodine is formed at the anode and dissolves in the potassium iodide solution, while iodine is reduced to iodide ions at the cathode. The gain and loss of iodine are usually determined volumetrically. However, when the quantity of electricity is very small, a colorimetric method for the estimation of iodine is more sensitive (1.316×10^{-3} g of iodine is liberated by 1 C).

The copper coulometer consists of a copper anode and copper cathode immersed in a solution of copper sulphate containing sulphuric acid and ethanol (to inhibit the oxidation of freshly deposited copper on the cathode). The cathode is weighed before and after electrolysis, and the quantity of electricity passed is calculated from the increased weight of the cathode (3.295×10^{-4} g of copper is deposited by 1 C).

In the water coulometer, electrolysis liberates oxygen and hydrogen from a pair of platinum electrodes placed close together in dilute sodium sulphate solution. If the total volume of gas discharged (corrected to STP) is V cm^3, the total quantity of electricity passed $= V/0.1741$ C.

Coulometric titrations. Coulometric methods of analysis are based on **Faraday's laws** of electrochemical equivalence. In normal volumetric analysis, physical methods (e.g., potentiometric, conductimetric, amperometric) are

66 Coulometric titrations

used to locate the end-point of a titration (*see* **Electrometric titrations**); in such titrations one of the reagents is a standard. In contrast, coulometric titrations require no standard solutions because a definite amount of reagent is generated in the solution; this amount is determined directly by the number of coulombs passed during the titration. The equivalence point is detected by any of the usual indicator or other instrumental methods.

Thus, in the titration of a base, water is used for the generation of protons in solution; these immediately react with the base. At the end-point, detected by an indicator, the total amount of electricity passed is determined, and hence the amount of protons. An accurate knowledge of the current passed and the time elapsed gives the number of coulombs passed; a simpler procedure is to pass a current, kept as constant as possible, through the titration cell and a silver **coulometer** in series.

This method has many distinct advantages. It requires no standard solution; a single electrical standard replaces the large number of chemical standards generally employed. Unstable reagents, such as chlorine, can be generated and used immediately without fear of decomposition or change of composition. As there is no minimum to the magnitude of the current used to generate the reagents electrically, microadditions of reagents can be made; one of the main sources of error is in the accurate detection of the end-point. Since the titrant is actually the electrons themselves there is no dilution of the sample.

Table 10 illustrates the versatility of the coulometric technique. Figure 32 shows a block diagram for a coulometric titration apparatus; included is a

Table 10
Typical applications of coulometric titrations

Substance to be estimated	Titrant generated	Substance used for generation of titrant
Anodic generation		
Base	H_3O^+	H_2O
As^{3+}, SO_3^{2-}	Cl_2	Cl^-
As^{3+}, Sb^{3+}, U^{4+}, I^-, SCN^-, NH_2OH, phenol, aniline, resorcinol, mustard gas	Br_2	Br^-
As^{3+}, $S_2O_3^{2-}$, H_2S	I_2	I^-
Fe^{2+}, hydroquinone	Ce^{4+}	Ce^{3+}
Cl^-, Br^-, I^-	Ag^+	Ag
Cathodic generation		
Acid	OH^-	H_2O
MnO_4^-, CrO_4^-, Br_2, Cl_2	Fe^{2+}	Fe^{3+}
CrO_4^-, VO_3^-, Br_2	Cu^+	Cu^{2+}

Coulometry

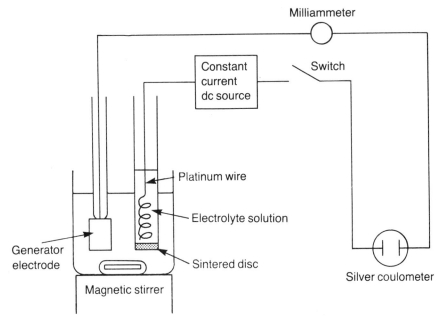

Figure 32. Apparatus for coulometric titration

constant current dc source and a silver coulometer. The silver coulometer can be replaced by an accurate milliammeter and an electric timer. During the early phases of a titration, the switch is kept in the closed position for extended periods. As the end-point is approached small additions of the 'reagent' are made by closing the switch for shorter and shorter periods. See James and Prichard (1977).

Further reading
D. A. Skoog and D. M. West, *Fundamentals of Analytical Chemistry*, 3rd edn, Holt, Reinhart and Winston, New York (1976)

Coulometry. A method of analysis which depends on **Faraday's laws** of proportionality. The measurement of the quantity of electricity, or the time for which a constant current passes, can be used for analytical determinations. The method requires carefully controlled conditions, the **current efficiency** must be 100 percent and the end-point accurately determined.

In the determination of the thickness of an electrodeposit a known small area is used as the anode, while the rest of the article is blocked off with a suitable coating. The electrodeposit is dissolved at an anode potential which leaves the base metal unattacked; the sudden rise in potential indicates the end of the process (i.e. the dissolution of the deposit). The quantity of electricity used is measured with a sensitive **coulometer** in series with the cell.

Coulometry can be applied to the quantitative determination of ions in solution—*electrogravimetry*—using the mass of metal deposited on the cathode. When two or more metal ions are present, the electrolysis must be conducted at a controlled potential to obtain selective deposition, since each metal is discharged at a potential near to its electrode potential.

Mixtures can also be analysed by *coulogravimetry*, in which the total mass of metal deposit ($w = w_1 + w_2$) and the quantity of electricity Q are measured.

$$Q = F\left(\frac{w_1 z_1}{A_1} + \frac{w_2 z_2}{A_2}\right)$$

where A_1 and A_2, and z_1 and z_2 are the relative atomic masses and valence, respectively, of the two metals. These simultaneous equations can be solved to determine w_1 and w_2.

Current (I; dimensions: $\varepsilon^{1/2} m^{1/2} l^{3/2} t^{-2}$; units: A = C s^{-1}). The rate of transfer of electricity. The practical unit of current is the ampere, that is the transfer of 1 coulomb per second. The ampere is one-tenth of the electromagnetic unit of current. The international ampere is the current which, when passed through a solution of silver nitrate, deposits metallic silver at a rate of $1.118\,00 \times 10^{-3}$ gram per second. *See also* **Electric units**.

Current distribution. In any electrochemical process, it is important to know that the current passing through an electrode does so uniformly, and that no part of the electrode is either unused or is passing a current density greater than it may safely do. The current distribution is discussed in terms of increasingly complex factors by which it is determined. The primary current distribution is obtained from simple geometrical considerations and a knowledge of the **potential distribution**. The inclusion of activation control of the reaction rate (*see* **Overpotential**) tends to smooth out the current distribution and leads to the secondary current distribution. Finally, changes in concentration of the reactants across the electrode surface may be important at high current density (*see* **Diffusion in electrolyte solutions**). These effects are included in the tertiary current distribution.

Current efficiency. The percentage of the current used in the electrolytic process under consideration, that is

$$\text{current efficiency} = \frac{\text{actual yield of element or compound}}{\text{theoretical yield (Faraday's laws)}} \times 100$$

The deposition of foreign elements, evolution of extraneous gases, mechanical loss, vaporization of deposit at high temperatures, side reactions and leakage of current may all contribute to the reduction of the current efficiency.

Cyclic voltammetry. *See* **Linear sweep voltammetry**.

D

Daniell cell. An electrochemical cell consisting of a zinc electrode dipping into a zinc sulphate solution and a copper electrode dipping in a copper sulphate solution; the two solutions are separated by a permeable membrane such as a sintered glass disc

$$\ominus \ \text{Zn} \ | \ \text{Zn}^{2+} \ \vdots \ \text{Cu}^{2+} \ | \ \text{Cu} \ \oplus$$

When the external circuit is completed, the copper electrode is positive, and positive current flows from left to right inside the cell and from right to left outside. Electron flow through the external circuit is left to right. The electrode reactions are

$$\text{Zn} \rightarrow \text{Zn}^{2+} + 2e$$
$$\text{Cu}^{2+} + 2e \rightarrow \text{Cu}$$

and the cell reaction

$$\text{Zn} + \text{Cu}^{2+} \rightarrow \text{Zn}^{2+} + \text{Cu}$$

Although both electrodes are reversible, diffusion of ions will occur at the liquid junction and the e.m.f. will vary continuously with time. The liquid junction and the corresponding **liquid junction potential** can be almost eliminated by interposing a **salt bridge** between the two solutions.

The theoretical standard e.m.f. of this cell $E^\ominus = 0.339 - (-0.761) = 1.100$ V; hence the standard free energy change for the reaction $\Delta G^\ominus = -2 \times 96\,500 \times 1.100 = -212$ kJ. The value of K_{therm} calculated from the standard free energy change is 1.6×10^{37}, indicating complete replacement of copper ions by metallic zinc. *See also* **Thermodynamics of cells.**

Debye–Falkenhagen effect. *See* **Conductance at high frequencies.**

Debye–Hückel activity equation. A theoretical equation relating the activity coefficient of an ion in an electrolyte to the ionic strength of the solution. The main assumptions made in its derivation are: (1) the spherically symmetrical Poisson equation is combined with the Boltzmann equation and linearized; (2) departures from ideality are due solely to coulombic interactions; and (3) the bulk relative permittivity is used.

70 Debye–Hückel–Onsager theory

$$\log \gamma_i = \frac{-A|z_i^2|I^{1/2}}{1 + B\mathring{a}I^{1/2}} \quad (41)$$

The second term in the denominator, in which $\mathring{a}$ is the effective diameter of the ion, makes allowance for the finite size of the ion. As $I \to 0$, the equation reduces to the limiting form

$$\log \gamma_i = -A|z_i^2|I^{1/2}$$

The corresponding equations for the mean ionic activity coefficient of an electrolyte are

$$\log \gamma_\pm = \frac{-A|z_+||z_-|I^{1/2}}{1 + B\mathring{a}I^{1/2}} \quad (42)$$

and

$$\log \gamma_\pm = -A|z_+||z_-|I^{1/2}$$

where

$$A = \frac{e^3}{2.303 \times 4\pi} \left(\frac{N_A}{2 \times 10^3}\right) (\varepsilon_0 \varepsilon_r kT)^{-3/2}$$

and

$$B = \left(\frac{2N_A e^2}{10^3 \varepsilon kT}\right)^{1/2}$$

For aqueous solutions at 298 K, $A = 0.509$ mol$^{-1/2}$ kg$^{1/2}$ and $B = 3.291 \times 10^9$ m^{-1} mol$^{-1/2}$ kg$^{1/2}$. The B term in the denominator indicates that the log $\gamma_\pm/I^{1/2}$ curves should diverge upwards from the limiting law as the concentration increases (see Fig. 4). The limiting equation gives values of $\gamma_\pm$ in excellent agreement with experimental values up to $I = 0.01$, but large deviations are observed even at this ionic strength if the product $|z_+||z_-| \geq 4$.

Debye–Hückel–Onsager theory. *See* **Conductance equations.**

Demineralization of water. *See* **Electrodialysis.**

Depth of discharge (DOD). The percentage rated capacity removed from a battery during discharge: for example, a 200 A h battery is 40 percent discharged when its capacity has fallen to 120 A h.

Desalination. *See* **Electrodialysis.**

Dielectric constant. *See* **Permittivity.**

Diffusion in electrolyte solutions. Ions move in electrolyte solutions under the influence of the electric field. Neutral molecules and ions may also move if a concentration gradient exists in the solution. In any reaction which consumes or liberates a soluble species, that species will diffuse towards or away from the electrode surface. The gradient of the chemical potential $(\partial \mu / \partial x)_t$ in the solution has units of $J\ m^{-1}$, and may be considered as the driving force for diffusion. **Fick's laws of diffusion** are followed. If the boundary conditions are known, the differential equations may be solved for simple systems to obtain the concentration of the reacting species as a function of time and distance from the electrode.

Diffusion near the electrode surface gives rise to concentration overpotential (*see* **Overpotential**), which in turn leads to the limiting current density (*see* **Diffusion-limited current**).

The diffusion coefficient is related to ion transport quantities by equating the gradient of the chemical potential with the electrical force on the ions. The Einstein equation relates the diffusion coefficient D_i to the ionic mobility u_i

$$D_i = \frac{u_i kT}{|z_i|e}$$

where z_i is the charge on the ion. The Nernst–Einstein equation is derived from the above equation. For a 1:1 electrolyte (i.e. one having equal numbers of anions and cations per mole)

$$\Lambda = \frac{z^2 F^2}{RT}(D_+ + D_-)$$

By applying the Stokes formula relating viscosity to mobility to the Einstein equation, the Stokes–Einstein equation is

$$D_i = \frac{kT}{6\pi \eta a}$$

where a is the radius of the ion.

Diffusion-limited current. In an electrolytic process, ions are used up or are generated at the electrodes leading to concentration changes at the outer Helmholtz plane (OHP). Consider a reaction in which a reactant is consumed at the electrode (e.g., metal plating). For a steady-state current, the concentration of reactant near the electrode surface is given by the solution of **Fick's laws of diffusion** and is illustrated in Figure 33. The flux (J) of a reactant reaching the OHP is related to the current density and concentration gradient by

$$J = \frac{i}{nF} = -D\left(\frac{\partial c}{\partial x}\right)_{x=0} \tag{43}$$

72 Diffusion-limited current

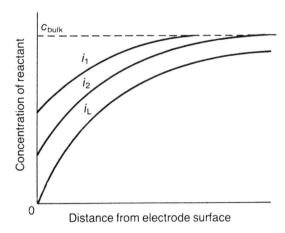

Figure 33. Concentration of a reactant near an electrode surface where $i_1 < i_2 < i_L$

Increasing currents may be sustained by increasing the concentration gradient until the concentration at the electrode falls to zero. At this point, the limiting current is reached. For many systems, a complete analytical solution of the differential equations is not possible. A simplification may be made by assuming that the change in concentration near the electrode is confined to a diffusion layer of thickness δ, in which there is a uniform concentration gradient (*see* Figure 34); δ is called the Nernst diffusion layer thickness. Eq. (43) becomes

$$J = \frac{i}{nF} = -D \frac{c_{\text{bulk}} - c_{x=0}}{\delta} \qquad (44)$$

At the limiting current density (i_L), $c_{x=0} = 0$ and thus

$$i_L = -\frac{nFDc_{\text{bulk}}}{\delta} \qquad (45)$$

δ is of the order of 0.5 mm in unstirred solutions, decreasing to around 0.01 mm in well-stirred solutions. Any attempt to increase the current beyond i_L results in a rapid increase in potential. The time taken for this to happen under constant current conditions is the **transition time**. Measurement of i_L when δ is controlled, as in **polarography** or at a **rotating-disc electrode**, allows determination of the bulk concentration.

Effect of limiting current density on current–voltage characteristics. The influence of the diffusion-limited current on the **Butler–Volmer equation** stems from the difference in concentration of reactant or product at the OHP compared with that in the bulk of the solution. For the general reaction

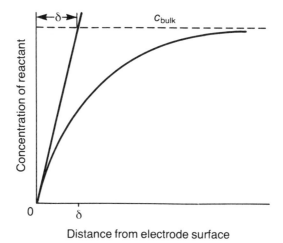

Figure 34. Linearized concentration gradient at the diffusion-limited current showing the Nernst diffusion layer thickness δ

$$ox^z + e \rightarrow red^{z-1}$$

$$a_{ox}^{OHP} = a_{ox}^{bulk}\left(1 - \frac{i}{i_{c,L}}\right)$$

$$a_{red}^{OHP} = a_{red}^{bulk}\left(1 - \frac{i}{i_{a,L}}\right)$$

where $i_{a,L}$ and $i_{c,L}$ are the anodic and cathodic limiting current densities, respectively. The Butler–Volmer equation is amended accordingly

$$i = i_0\left\{\left(1 - \frac{i}{i_{c,L}}\right)\exp\left(-\frac{\beta F}{RT}\eta\right) - \left(1 + \frac{i}{i_{a,L}}\right)\exp\left[\frac{(1-\beta)F}{RT}\eta\right]\right\}$$

In the Tafel region (*see* **Tafel equation**)

$$\eta_c = \frac{RT}{\beta F}\ln\frac{i_0}{i_{c,L}} - \frac{RT}{\beta F}\ln\frac{i/i_{c,L}}{1 - i/i_{c,L}}$$

$$\eta_a = -\frac{RT}{(1-\beta)F}\ln\frac{i_0}{i_{a,L}} + \frac{RT}{(1-\beta)F}\ln\left(-\frac{i/i_{a,L}}{1 + i/i_{a,L}}\right)$$

Dimensionally stable anode (DSA). The production of chlorine in the diaphragm cell requires an anode that is inert in the extremely corrosive environment of hot chloride at anodic potentials (*see* **Electrolysis of brine**). Graphite is the traditional material; it has a life of some 12 months during which two 36-kg anodes can corrode away. In 1965 a metal anode consisting

74 Dipole potential

of titanium covered with precious metal oxides (platinum (IV) oxide or ruthenium dioxide) was patented. The dimensionally stable anode is lighter, operates at a lower overpotential and is corroded at a minimal rate giving lifetimes of ten years or more in a diaphragm cell. These anodes operate at twice the current density and have a better **current efficiency** than graphite. Other dimensionally stable anodes have been constructed on a tantalum base. The dimensionally stable anode is also known as a precious metal anode (PMA), noble metal-coated titanium anode (NMT) or platinum–titanium anode (PTA). See Kirk-Othmer (1981).

Dipole potential. *See* **Surface potential.**

Dissociation constant (K_a; units: mol dm^{-3}). For a weak acid HA

$$HA + H_2O \rightleftharpoons H_3O^+ + A^-$$

the dissociation constant or ionization constant is given by

$$K_a = \frac{a_{H3O+} a_{A-}}{a_{HA}} = K_c \gamma_{H3O+} \gamma_{A-} = K_c \gamma_\pm^2 \qquad (45)$$

and $pK_a = -\log K_a$.

Weak bases can be treated in a similar manner.

Determination of dissociation constants.
1) Titration curves (*see* Figure 35) (*see also* **Electrometric titrations**) only give the classical dissociation constant, K_c; a correction may be applied using the **Debye–Hückel activity equation** to give K_a. After the addition of t cm^3 of titrant, the concentration of free acid is proportional to $T - t$, where T is the volume of titrant at the equivalence point, and the concentration of salt is proportional to t. Hence, using the Henderson equation

$$pH = pK_c + \log \frac{c_{A-}}{c_{HA}} = pK_c + \log \frac{t}{T-t} \qquad (46)$$

The graph of pH against $\log t/(T-t)$ is linear and of intercept pK_c (*see* Figure 36). At the point of 'half titration', $t = T/2$; hence, $pH = pK_c$.

2) Conductometric method. From the measured **conductance** of a weak acid at low concentrations, the degree of dissociation α ($=\Lambda_c/\Lambda_c'$) and K ($=\alpha^2 c/(1-\alpha)$) can be calculated. Λ_c', the molar conductivity which the electrolyte would have if it were fully dissociated at concentration c, may be calculated using the Kohlrausch's law of independent migration of ions

$$\Lambda_{c,HA}' = \Lambda_{c,HCl} + \Lambda_{c,NaA} - \Lambda_{c,NaCl}$$

where the Λ_c values for the strong electrolytes all refer to the same concentration c.

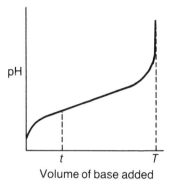

Figure 35. Typical titration curve

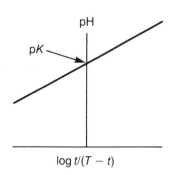

Figure 36. Graph of pH against $\log t/(T-t)$

Since

$$\log K_c = \log K_a + 2A(\alpha c)^{1/2} \qquad (47)$$

the graph of $\log K_c$ against $(\alpha c)^{1/2}$ is linear and of intercept $\log K_a$.

3) E.m.f. method using **concentrations cells** without a liquid junction (*see* **liquid junction potential**): for example

$$\ominus \text{ Pt,H}_2(g) \mid \text{HA, NaA, NaCl} \mid \text{AgCl(s),Ag} \oplus$$
$$(m_1) \quad (m_2) \quad (m_3)$$

for which the overall cell reaction is

$$\tfrac{1}{2}\text{H}_2(g) + \text{AgCl(s)} \to \text{Ag(s)} + \text{H}^+ + \text{Cl}^-$$

and the e.m.f. is

$$E = E^\ominus_{\text{AgCl,Ag,Cl}^-} - \frac{RT}{F} \ln a_{\text{H}^+} a_{\text{Cl}^-} \qquad (48)$$

The hydrogen ions are provided by the dissociation of the weak acid; substituting for a_{H^+} from Eq. (45) in Eq. (48) gives

$$\frac{F(E - E^\ominus_{\text{AgCl,Ag,Cl}^-})}{2.303 RT} + \log \frac{m_{\text{HA}} m_{\text{Cl}^-}}{m_{\text{A}^-}} = -\log \frac{\gamma_{\text{HA}} \gamma_{\text{Cl}^-}}{\gamma_{\text{A}^-}} - \log K_a \qquad (49)$$

The left-hand side of Eq. (49), calculated from the measured values of E at various values of m_1, m_2 and m_3, is plotted against $I^{1/2}$; the intercept of the resulting line is $-\log K_a$. Since HA is partially dissociated, it follows that $m_{\text{HA}} = m_1 - m_{\text{H}^+}$; hence m_{H^+} is required for the calculation. This can be obtained from an assumed approximate value of K_a, and by a series of successive approximations, a good value of K_a may be obtained. It is usual to maintain m_1/m_2 at unity and vary m_3.

4) **Spectrophotometric method.** The absorbance of a fixed concentration of the acid in a series of buffer solutions of known pH is measured at the wavelength of maximum absorbance. If ε_{HA} and ε_{A^-} are the molecular extinction coefficients (i.e. A/cl, where A is the absorbance and l the path length) of HA and A$^-$, respectively, ε the measured extinction coefficient, and c/mol dm^{-3} the total acid concentration, then

$$\varepsilon c = \varepsilon_{HA} c_{HA} + \varepsilon_{A^-} c_{A^-} \quad (\text{H}_3\text{O}^+ \text{ does not absorb})$$

and

$$c = c_{HA} + c_{A^-}$$

Thus

$$\text{pH} = \text{p}K_a + \log \frac{c_{A^-}}{c_{HA}} + \log \gamma_{A^-}$$

$$= \text{p}K_a + \log \left(\frac{\varepsilon_{HA} - \varepsilon}{\varepsilon - \varepsilon_{A^-}} \right) - \frac{A|z|^2 I^{1/2}}{1 + 1.25 I^{1/2}}$$

Hence, the graph of pH against $\log [(\varepsilon_{HA} - \varepsilon)/(\varepsilon - \varepsilon_{A^-})]$ is linear and of intercept $\text{p}K_a - A|z|^2 I^{1/2}/(1 + 1.25 I^{1/2})$.

A decrease in the relative permittivity (*see* **Permittivity**) of the solvent causes an increase in the electrostatic forces between anions and cations, and hence an increase in the formation of covalent bonds. This is accompanied by a decrease in the dissociation constant of a weak acid dissolved in it. For acetic acid at 298 K in water ($\varepsilon_r = 78.5$), $K_a = 1.754 \times 10^{-5}$ mol dm^{-3}, whereas in 82 percent aqueous dioxane ($\varepsilon_r = 9.5$), $K_a = 3.1 \times 10^{-11}$ mol dm^{-3}.

The dissociation constant of a weak acid varies with the temperature according to an equation of the type

$$2.303 R \log K_a = -\frac{A}{T} + C - DT$$

where A, C and D are constants. K_a passes through a maximum value at a temperature of $(A/D)^{1/2}$, which is about room temperature for most weak acids. The maximum value is given by

$$2.303 R \log K_a = C - 2(AD)^{1/2}$$

See James and Prichard (1977); Robinson and Stokes (1965).

Donnan membrane equilibrium. The distribution of simple electrolytes on the two sides of a membrane, freely permeable to these electrolytes but impermeable to colloidal ions, in the presence of a colloidal electrolyte on one side of the membrane. If an aqueous solution of the sodium salt of a protein Na$_z$R (where R is a colloidal anion) of concentration c_1 is separated by such a membrane from an equal volume of sodium chloride of concentration c_2, then some sodium chloride, x, passes through the membrane until equilibrium is established, when the **chemical potential** of sodium chloride is the same on both sides.

Initially

Na^+	zc_1	Na^+	c_2
R^-	c_1	Cl^-	c_2

At equilibrium

Na^+	$zc_1 + x$	Na^+	$c_2 - x$
R^-	c_1		
Cl^-	x	Cl^-	$c_2 - x$

This condition is fulfilled when

$$a_{Na^+,1} a_{Cl^-,1} = a_{Na^+,2} a_{Cl^-,2}$$

assuming that in dilute solutions the activity coefficients are unity

$$(zc_1 + x)x = (c_2 - x)^2$$

that is

$$\frac{x}{c_2} = \frac{c_2}{zc_1 + 2c_2}$$

The fraction of sodium chloride diffusing through the membrane (x/c_2) is the smaller, the greater the concentration of R^- and the greater its valence. If there is only a small amount of Na_zR present, then the distribution of sodium chloride (($c_2 - x$)/x) on the two sides of the membrane is nearly one, but if $c_1 \gg c_2$, nearly all the sodium chloride remains on the other side of the membrane to R^-.

At equilibrium there is an electrical potential difference, the membrane potential, caused by the establishment of a **concentration cell**, given by

$$E = \frac{2.303 RT}{F} \log \frac{c_{NaCl,RH}}{c_{NaCl,LH}} = \frac{2.303 RT}{F} \log \frac{c_2 - x}{x}$$

$$= \frac{2.303 RT}{F} \log \left(1 + \frac{zc_1}{c_2}\right)$$

With a colloidal anion, E is positive; with a colloidal cation, E is negative. With a polyvalent ion ($z > 1$), values of E are larger than in Figure 37 (e.g., for $z = 10$, $c_1/c_2 = 100$, $E = 0.178$ V).

Dorn effect. One of the electrokinetic phenomena (see **Electrokinesis**); it is the opposite of **electrophoresis**. Charged particles falling through a stationary

78 Double layer

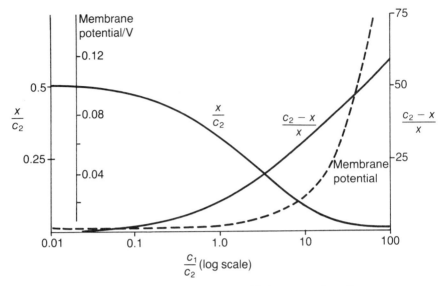

Figure 37. Donnan membrane equilibrium curves ($z = 1$)

liquid set up a potential difference in the liquid column, opposing the motion of the particles. Large effects (up to 100 V) have been measured, but accurate results are more difficult to obtain than for the other related effects. The potential generated, often known as the sedimentation potential, must be taken into account in the measurement of sedimentation rates in the ultracentrifuge.

Double layer. *See* **Electrical double layer.**

Down's cell. *See* **Sodium electrometallurgy.**

Dropping mercury electrode. *See* **Polarography.**

Dry cell. *See* **Leclanché cell.**

Dynamic hydrogen electrode (DHE). A reference electrode in acid or alkaline solutions which does not need a supply of hydrogen gas. The dynamic hydrogen electrode consists of an electrolysis cell of two platinum black electrodes through which a small (approximately 2 mA) current passes. The great activity of platinum towards hydrogen evolution ensures that the **overpotential** at the cathode is only a few millivolts, and so the potential at the cathode is near that of the reversible hydrogen electrode at the pH of the solution. The reference potential is affected by reducible species in solution or those (e.g., chloride ions) which adsorb at platinum surfaces.

E

Electrical double layer. At any boundary between two phases the intermolecular forces will be different from those in the interior of either phase. Consequently, the concentrations of any mobile species are likely to differ in this region from those in the bulk phases, and any dipolar molecules in the region may be oriented by surface forces. A result to be expected is the setting up of an electrical double layer at the surface, one side of which carries a positive charge and the other an equal negative charge.

The separation of charges may arise in various ways. At a solid/electrolyte solution interface, there is likely to be preferential attraction of one ion to the solid surface, with a corresponding excess of the counter ion in the adjacent solution. (Even at the air surface of an electrolyte solution there is an electrical separation, as most anions tend to approach the surface more closely than cations.) Above all, the effect will be important where a charged species is able to pass across the boundary from one phase to the other, as in the ionization of surface groups of a colloidal particle, or at the interface salt crystal/solution, or at a metal electrode.

The simplest model of an electrical double layer (given by Helmholtz and Perrin) is one in which a charge q is uniformly distributed over a plane or surface area A and is separated by a distance d from a similar plane with a charge $-q$. In the case of the electrode/electrolyte interface, the locus of the centres of the plane of ions in the solution is called the outer Helmholtz plane (OHP). Writing $\sigma = q/A$ for the surface charge density, we obtain the potential difference across the layer from the formula for a parallel plate capacitor

$$V = \frac{\sigma d}{\varepsilon_0 \varepsilon} \tag{50}$$

The **capacitance** of the layer is given by

$$C = \frac{\sigma}{V} = \frac{\varepsilon_0 \varepsilon}{d} \tag{51}$$

where C is in F m^{-2} if V is in volts and σ in C m^{-2}. Note that Eq. (51) predicts that the capacitance of the double layer is independent of potential. In 1905, Goüy pointed out that the Helmholtz double layer cannot apply when one side of the layer consists of mobile ions whose thermal energy would oppose their

80 Electrical double layer

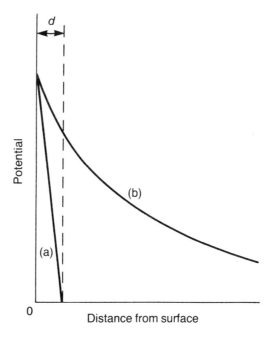

Figure 38. Electrical double layer: (a) Helmholtz double layer; (b) diffuse double layer

ordering by electrostatic forces. An equilibrium must result in which the excess of cations or anions will be at a maximum close to the surface, but will diminish gradually at increasing distances. The quantitative theory is exactly similar to the Debye–Hückel treatment of the atmosphere around a charged ion (*see* **Debye–Hückel activity equation**). This is the Goüy–Chapman model of a diffuse double layer; it is compared with the Helmholtz–Perrin model in Figure 38. The diffuse double layer is electrically equivalent to a charge, equal and opposite in sign to the net charge on the fixed side of the layer, spread out at a distance κ^{-1} from the surface and depends on the concentrations and valencies of all the ions present, and κ^{-1} is the thickness of the ionic atmosphere or Debye length. Hence the diffuse double layer can be treated as a parallel plate capacitor in estimating its effect, with κ^{-1} replacing d in Eqs. (50) and (51).

Later, in 1924, Stern proposed a synthesis of these two models. It is reasonable to suppose that some of the charge carriers in solution will be strongly attracted by surface forces, while the remainder will distribute themselves in a diffuse double layer. There are, therefore, two regions of charge separation: the first in which occurs a linear fall in potential from the electrode to the OHP and the second in which an exponential fall in potential results. If the charge on the OHP is less than that of the electrode, the

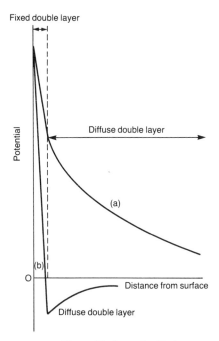

Figure 39. Stern double layer

remaining charge is balanced by the diffuse double layer. The charge on the OHP may, however, be greater than that of the electrode, leading to a reversal in sign of the diffuse layer. The two possibilities are illustrated in Figure 39. The capacity of metal/solution interfaces can be calculated from electrocapillary curves (*see* **Electrocapillary phenomena**), or from the growth of the potential during the passage of a very small charging current (*see* **Capacitance**). The data show that the Helmholtz and Goüy models are both unsatisfactory, but that the Stern model can form the basis for an explanation of the main experimental features. The picture of a metal/solution interface that can be built up from this evidence is somewhat as follows (*see* Figure 40). The layer in direct contact with the metal will consist of water molecules with their dipoles partly oriented to the metal surface; the direction and degree of orientation will depend on the charge carried by the metal. Interspersed among these may be some ions; these will be ions, usually anions, with a relatively low hydration energy or a strong adsorptive affinity for the metal, which therefore have a net tendency to exchange a water molecule for a metal atom as neighbour. This phenomenon is known as specific or contact adsorption, and together with the ions in the OHP and diffuse layer, the contact-adsorbed ions constitute a triple layer. There will also be an adsorption into this layer of any organic solute present, either because of specific adsorption

82 Electric units

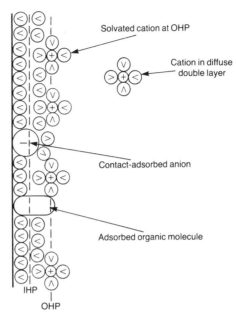

Figure 40. Possible structure of the electrode/electrolyte interface

forces or simply because it will tend (as at an air interface) to be squeezed out of the aqueous medium. The charge associated with this layer is said to reside at the inner Helmholtz plane (IHP). Strongly hydrated ions will be unable to approach the surface so closely, and will be separated from it by one or two water molecules. They are adsorbed at the OHP, and beyond this will be the diffuse double layer. At higher concentrations of electrolyte, the concentration of adsorbed ions increases and the diffuse double layer is compressed, so that its equivalent net charge is closer to the surface. The properties of the interface then approximate more closely to those of a Helmholtz double layer. At very low concentrations, the diffuse part of the double layer becomes relatively more important. See Eyring (1970); Fried (1973); Parsons (1961).

Electric units. In the electrostatic system of units, the unit electric charge Q is defined as that charge which, when placed 1 centimetre from an identical charge in a vacuum, repels it with a force of 1 dyne. When placed in an electric field, this unit charge will experience a force of E dynes tending to move it in the direction of the field, and E then measures the electric field strength in electrostatic units (e.s.u.). Similarly, the potential difference between two points will be 1 e.s.u. if 1 erg of work is performed when unit charge is moved from one point to the other. An alternative system of electromagnetic units (e.m.u.) is based on Ampere's theorem, which defines current in terms of its magnetic effect. In the SI system of units, the basic electric unit is the **ampere**

Table 11

Unit	e.m.u.	e.s.u.	Abs./Int.[a]
1 A	1×10^{-1}	3×10^9	0.999 835
1 C	1×10^{-1}	3×10^9	0.999 835
1 V	1×10^8	1/300	1.000 330
1 Ω	1×10^9	$1/(9 \times 10^{11})$	1.000 495
1 F	1×10^{-9}	9×10^{11}	0.999 505

[a] Factors by which the 'international unit' must be multiplied to convert it into the absolute unit now employed.

(A). The other derived electric units are: the **coulomb** (C = A s); the **volt** (V = J A^{-1} s^{-1}); the **ohm** (Ω = V A^{-1}); and the **farad** (F = C V^{-1}). Conversion factors for these quantities are given in Table 11. The 'international units' were practical units adopted at an international conference in 1918 and much of the data in the literature are given in these units. They were replaced by the absolute units in 1946.

Electrocapillary phenomena. The interfacial tension (γ) of a mercury/electrolyte interface varies with the potential of the mercury. When a potential is applied to mercury in a Lippmann electrometer (*see* Figure 41), the position

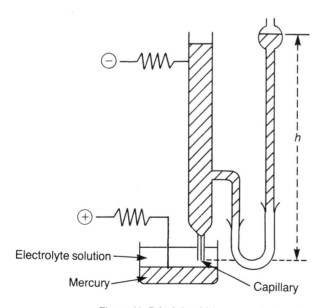

Figure 41. Principle of Lippmann electrometer

84 Electrocapillary phenomena

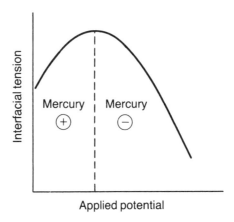

Figure 42. Schematic variation of interfacial tension with applied potential (electrocapillary curve) (Eq. 54)

of the mercury meniscus changes; the meniscus may be returned to its original position by adjusting the height (h) of the mercury reservoir. The amount by which the height is changed is a function of the change in the interfacial tension caused by the applied potential. Initially the mercury is positively charged with respect to the solution, but as it is made more negative the interfacial tension at first increases, reaches a maximum value and then decreases (*see* Figure 42).

The Lippmann equation

$$\left(\frac{\partial \gamma}{\partial E}\right)_{T,P,\mu} = -\sigma \qquad (52)$$

expresses the variation of the surface tension with the applied potential (E). Writing the surface charge density, $\sigma = CE$ (where C is the capacitance of the double layer regarded as a condenser), then at constant T, P, μ

$$\partial \gamma = - CE \, \partial E \qquad (53)$$

which on integration gives

$$\gamma = -\tfrac{1}{2}CE^2 + \text{constant} \qquad (54)$$

This is the equation to a parabola, at the maximum $\partial \gamma / \partial E = 0$, $E = 0$ and $\sigma = 0$ (*see* **Potential of zero charge**); that is the electrical double layer is discharged. The capacity of the double layer at the mercury/electrolyte interface may be obtained by differentiating Eq. (52) to give

$$\left(\frac{\partial^2 \gamma}{\partial E^2}\right)_{T,P,\mu} = -\frac{\partial \sigma}{\partial E} = C$$

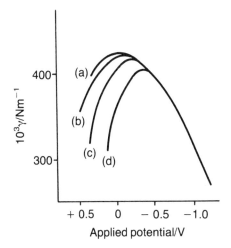

Figure 43. Interfacial tension between mercury and 1.0 mol dm^{-3} solutions at various potentials. (a) Potassium hydroxide, (b) sodium chloride, (c) sodium bromide and (d) potassium iodide

Electrocapillary curves for different electrolytes (except for potassium nitrate over a limited concentration range) are not normally parabolic but are distorted. If the capacity of the double layer remained constant for a given electrode, identical electrocapillary curves would be recorded irrespective of the nature of the electrolyte. Alkali metal nitrates show almost coincident parabolas, but other salts of the alkali metals each give their own characteristic curve (*see* Figure 43).

Large negative ions are strongly adsorbed at the interface in the order $S^{2-} > I^- > CNS^- > Br^- > Cl^- > OH^- > F^-$; the adsorption is enhanced when the applied potential renders the mercury positive. Similar effects of the sequence of cations $NEt_4^+ > NMe_4^+ > Tl^+ > Cs^+ > Na^+$ are evident in the region where negative potentials are applied to the mercury.

The water/mercury interface is widely used for the study of biological redox reactions; electron transfer is accompanied by a heterogeneous process between the adsorbed biological macromolecule and the electrode. Mercury electrodes may be used to estimate small amounts of protein and the detection of denaturation processes.

Electrocatalysis. A catalyst is a substance that alters the rate of a chemical reaction without suffering a change itself. In heterogeneous reactions this is usually accomplished by adsorption at the surface of the catalyst followed by reaction via a path having a lower energy of activation. In electrochemical reactions the reactants come close to, or are adsorbed onto, the electrode surface, and so it is likely that the nature of the surface will have an effect on

86 Electrocatalysis

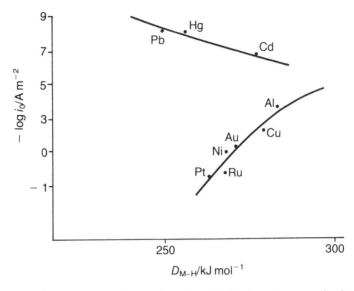

Figure 44. Exchange current density as a function of M–H adsorption energy for the evolution of hydrogen on a number of metals

the rate of the electrochemical reaction. This is seen in the variation by 10^{10} in the exchange current densities of the **hydrogen electrode reactions** on platinum ($i_0 = 10$ A m^{-2}) and on lead ($i_0 = 10^{-9}$ A m^{-2}). The **Butler–Volmer equation** contains non-potential-dependent terms which are taken into the **exchange current density**. A change in the electrode material can change the chemical activation free energy ($\Delta G^{\ddagger}$), which results in a change in i_0. It is expected, and found, that the exchange current density for a reaction without specific interactions between reactant and electrode (see **Electrical double layer**) should vary linearly with the work function of the electrode material. Another example of the effect of the chemical nature of the interaction between electrode and reactant is in the hydrogen electrode reaction, in which a correlation has been made between i_0 and the metal–hydrogen bond energy (see Figure 44). Differences in the extent of adsorption of reactants or intermediates may, by changes in concentration terms, lead to different kinetics. This behaviour may in turn be related to the d-band character of the metal. The reduction of oxygen (see **Oxygen electrode reactions**) on a range of platinum–gold alloys, in which the addition of gold fills up the d-band vacancies of platinum, shows a rise in activity with platinum content as an adsorbed intermediate M . . . OH$_2$ is stabilized.

In comparing electrocatalysts the choice of potential may be important. Figure 45 shows a **Tafel equation** plot for the oxidation of ethylene (ethene) on platinum and the alloy Pt$_{0.8}$Ru$_{0.2}$. The intersecting lines are reminiscent of

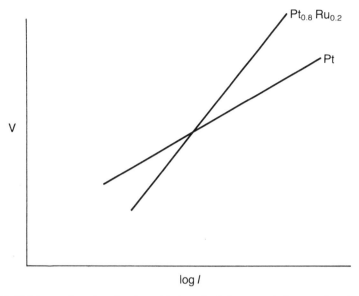

Figure 45. Tafel equation plots for the oxidation of ethylene on platinum and a platinum–ruthenium alloy

similar plots of rate against $1/T$ in heterogeneous chemical catalysis. Most comparisons are made between exchange current densities (i.e. when the **overpotential** is zero) but some authors favour comparison of currents at the **potential of zero charge** when the field across the surface is zero. However, at the potential of zero charge the free energy of the initial states of the electrons in the metal depends on the metal, which detracts from the use of this potential for comparison purposes.

Electrochemical energy conversion. Any thermodynamically spontaneous chemical reaction that may be written as the sum of two redox reactions may, in theory, yield the free energy of the reaction in the form of an electric current flowing through a potential difference given by $E = -\Delta G/nF$. Electrochemical energy convertors are divided into two main classes depending on whether the reactants are held within the device (battery) or are supplied externally (**fuel cell**).

The efficiency of an electrochemical convertor (i.e. the fraction of the energy of the reaction which may be usefully converted to electrical energy) is of importance. In electrochemical energy convertors, there are no Carnot cycle inefficiencies, however the ratio $\Delta G/\Delta H$ does take account of the entropy change of the reaction, and this is a measure of the intrinsic efficiency of the process. Many reactions involve gaseous reactants in which ΔS is negative, and thus the efficiency is less than 100 percent (although usually

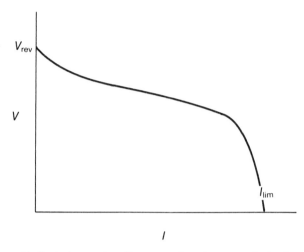

Figure 46. Typical current–voltage curve for a battery or fuel cell

greater than 80 percent). The true efficiency of a reaction also must include a component due to the practical voltage at which the device operates. This takes account of the **overpotential** at anode and cathode, and ohmic losses in the electrolyte.

$$\varepsilon = \frac{\Delta G\, V_{operating}}{\Delta H\, V_{rev}}$$

$$\varepsilon = \frac{V_{rev}(V_{rev} - \eta_c - \eta_a - IR)}{V_t V_{rev}} \tag{55}$$

Where V_{rev} is the reversible potential of the reaction, V_t is the **thermoneutral potential**, and η_a and η_c are the anodic and cathodic overpotentials, respectively.

The effect of the overpotential terms on the efficiency of the reaction is seen in the current–voltage curve of a cell. It starts from the theoretical maximum, which can be calculated from the free energy change of the reaction. At finite currents, it drops on account of the overpotential at the electrodes, and ultimately the efficiency (and voltage) falls to zero as the current approaches the limiting current density (*see* **Diffusion-limited current**) (*see* Figure 46). See also **Electrochemical storage**. See Bockris and Reddy (1973).

Electrochemical equivalent. *See* **Faraday's laws.**

Electrochemical machining. Utilization of the anodic dissolution of a metal to produce an article of a desired shape and size. It has advantages over the

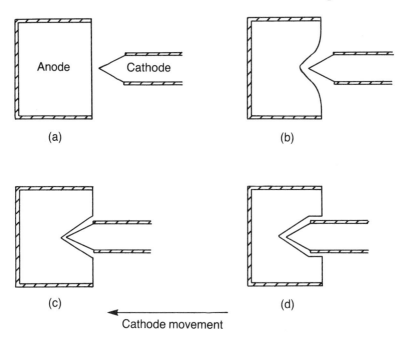

Figure 47. Progress of an electrochemical machining operation

normal mechanical processes for very tough metals or where mechanical stresses are to be avoided. Comparatively large amounts of metal are involved, and the electrochemical conditions used are similar, in some respects, to those of **electroforming**. Very high current densities (50–500 A cm^{-2}) are used at cell voltages of 5–25 V. An electrolyte of high conductance is circulated between the anode and cathode. Sodium chloride is commonly used, but tungsten may only be machined in strong alkali and sodium chlorate solution has been used for some steels.

A possible process is illustrated in Figure 47. The sides of the metal are insulated, for example, by a layer of plastic. The current density at the anode is highest opposite the tip of the cathode, and dissolution is most rapid here. The cathode is rigidly mounted and constantly advanced at a rate to keep the gap constant. This avoids sparking on the one hand, or a falling off of the current on the other. The required shape is eventually produced, being determined by the dimensions of the cathode and those of the (constant) gap between the electrodes.

Electrochemical potential ($\bar{\mu}$; units: J mol^{-1}). This is defined by the equation

$$\bar{\mu}_i = \mu_i + z_i F \phi = \mu_i^{\ominus} + RT \ln a_i + z_i F \phi \tag{56}$$

90 Electrochemical series

where $z_i F\phi$ is the work done in bringing a unit amount of charge $z_i e$ from infinity to a region where the Galvani potential is ϕ and μ_i is the **chemical potential** when $\phi = 0$. A positive ion ($z_i e$ positive) in a place with a positive potential ϕ will have an enhanced tendency to escape. If the potential is negative, the ion is stabilized and the escaping tendency is smaller.

In a system at equilibrium, the electrochemical potential of each species must remain constant throughout the system, if not the species i spontaneously diffuses from the region of higher to the region of lower electrochemical potential until equilibrium is established. If α and β represent two phases in equilibrium

$$\bar{\mu}_i^\alpha = \bar{\mu}_i^\beta$$

whence

$$\mu_i^\alpha + z_i F\phi_\alpha = \mu_i^\beta + z_i F\phi_\beta$$

The phase boundary potential $\Delta\phi_{\alpha\beta}$ is

$$\Delta\phi_{\alpha\beta} = \phi_\alpha - \phi_\beta = \frac{\mu_i^\alpha - \mu_i^\beta}{z_i F} \qquad (57)$$

Electrochemical series. A series of metals arranged in decreasing order of their tendency to pass into the ionic form by losing electrons; the most active metals are at the top and the least active at the bottom.

The series was first established experimentally by discovering which metal would displace others from their salts. Thus a strip of zinc immersed in a solution of copper sulphate is soon covered by a deposit of metallic copper while the zinc passes into solution. This is really an oxidation–reduction transfer of electrons

$$\text{Zn} \rightarrow \text{Zn}^{2+} + 2e$$
$$\text{Cu}^{2+} + 2e \rightarrow \text{Cu}$$
$$\overline{\text{Cu}^{2+} + \text{Zn} \rightarrow \text{Cu} + \text{Zn}^{2+}}$$

The tabulation of standard **electrode potential** (*see* Appendix, Table 3) permits the extension of the original series of metals to non-metallic elements.

Metals which liberate hydrogen gas from dilute acids are above hydrogen, with a negative electrode potential, whereas those metals and non-metallic elements which will not liberate hydrogen from dilute acids are below hydrogen in the series and have positive electrode potentials.

Electrochemical storage. Energy, stored in chemical bonds within a cell, which may be recovered as electricity. A primary cell contains the reactants of an exoenergetic reaction. When consumed the life of the cell is over. A great variety of chemical reactions can be used as the basis for primary cells. Most practical arrangements will give voltages in the range 1–2 V, corresponding to reactions with a free energy change of about 210 kJ mol^{-1}. This, for instance, is the value for the **Daniell cell** reaction, and the theoretical voltage is 1.10 V (*see* **Electrochemical energy conversion**).

The choice of reaction is based on practical considerations: cheapness of materials, a high **energy density** (W h kg^{-1}), high storage density (C kg^{-1}), high power density (W kg^{-1}) and a good shelf life (no deterioration when the cell is idle). One basis for comparison of different cells is the cell voltage–current diagram (*see* Figure 46). Another criterion of cell performance is the voltage–time curve (*see* Figure 6) when the cell is discharged continuously through a fixed resistance. A series of curves for different resistances may be constructed, and also curves for the intermittent use of the cell; the latter shows how well the cell recovers, through depolarizing effects and diffusion, after a period of high current withdrawal. In the alkaline cell, the products of electrolytic action are precipitated, and the conditions in the cell, and its voltage, remain very constant throughout its life.

A secondary cell, or accumulator, is a cell which can be recharged by passing a current through it from an external source. Many cell reactions that are chemically reversible prove unsuitable for this purpose because the charge–discharge cycle tends to bring about physical changes: for example, in the condition of the electrodes. The practical choice is also limited to systems needing only a single electrolyte. The **lead–acid battery** and the **nickel–cadmium battery** are among the most important secondary batteries. Others include the **silver–zinc battery**, the **metal–air battery** and the **alkaline cell**.

A secondary battery, like the primary battery, should have high energy density, power density and storage density, but must also have a good cycle life, that is the number of times the battery may be recharged after a certain degree of discharge, and a high energy efficiency, which is the ratio of the energy obtained from the battery to the energy required for charge. See Crompton (1982).

Electrode. A conductor, usually a metal, immersed in an electrolyte solution (i.e. an electronic conductor in contact with an electrolytic conductor) that serves either as a source or sink of electrons. The application of an e.m.f. between two such electrodes in an aqueous solution will give rise to electrolysis. Alternatively as a result of chemical reactions at the electrode surfaces an e.m.f. may be produced. Inert electrodes (e.g., platinum) serve only to transfer electrons to and from the solution; reacting electrodes enter chemically into the electrode reaction. The nature of the electrode material determines the electrochemical reactions that occur (*see* **Electrocatalysis**). *See also* **Half cell; Modified electrode.**

92 Electrodecantation

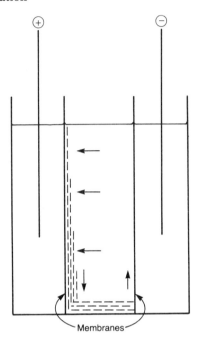

Figure 48. Schematic representation of an electrodecantation cell

Electrodecantation. (Electrogravitational separation with electrophoretic convection). A technique which is based on a stratification phenomenon that takes place when a colloidal suspension is subjected to an electrical field between vertical membranes which are impermeable to the colloid (*see* Figure 48). The charged particles migrate under the influence of the electrical field towards one of the electrodes but are retained and accumulated on the membrane. Under certain conditions, the thin concentrated layer on the membrane surface moves up or down between the membrane and the surrounding liquid according to its relative density; at the other membrane, the movement of the dilute layer is in the opposite direction.

The stratified layers can be separated by decantation. The separation into layers is independent of particle size. The behaviour of the material retained on the membrane depends on the relative values of the velocity of migration in the electric field (v_e) and the velocity of gravitational movement (v_g). If $v_g > v_e$, then stratification will occur continuously, but if $v_e > v_g$, the deposit will accumulate on the membrane. If the deposit is allowed to remain on the membrane, loss of liquid by electroosmosis will occur and the colloid will coagulate. Periodic reversal of the polarity of the electrodes will free the deposit from the membrane, and this immediately stratifies.

The technique is widely used in biological systems: for example, in the isolation of immunoglobulin antibody activity and the initial fractionation of enzymes from crude material. It is also employed industrially in the concentration of rubber latex.

Electrode kinetics. *See* **Butler–Volmer equation; Electrode reaction mechanisms.**

Electrodeposition of metals. Most metallic cations are readily discharged at potentials close to their equilibrium values, but with iron, nickel and other transition metals there is a significant activation overpotential (*see* **Overpotential**). The structure of the deposit produced depends greatly on the conditions of the electrolysis. When deposition takes place readily and is unimpeded, relatively large crystals tend to form, which may continue the crystalline structure of the underlying metal. Good stirring, elevated temperature and the use of a simple salt of the metal at a high concentration will favour this type of deposit. However, any constraint in the overall process tends to lead to a fine-grained microcrystalline deposit. When this is the objective, it is common to use a complex salt of the metal as electrolyte, to add to the bath nonelectrolytes that are strongly adsorbed at the surface, or to work at high current densities and overpotentials.

When the electrolyte contains more than one cation, their simultaneous discharge becomes possible. This will occur if their deposition potentials (under the conditions used) are close to one another, and it can lead to **alloy electrodeposition**. The most general case, however, is of course where hydrogen ion is the second cation, and the simultaneous evolution of hydrogen is a common accompaniment of metal deposition. The amount of current that will be dissipated in this way can be calculated if the current–potential curves for the two processes are known. It must be remembered that the curve for hydrogen evolution depends on the nature of the cathode surface, as well as on the pH of the solution. Figure 49 illustrates the deposition of zinc from a neutral solution of zinc sulphate. At the point at which the curves intersect the current efficiency for zinc deposition is 50 percent, but at high current densities it is much greater.

The mechanism of the discharge of a metal ion involves a number of stages, such as are shown in Figure 50. Initially, the metal cation is in solution and associated with a complete hydration shell. In its final state, it has attained a stable position in the metal lattice, but still has partial ionic character on account of the delocalized bonds associated with a metallic lattice. The intermediate stages are thought to involve separate steps in which the ion first becomes adsorbed on the surface as an 'ad-ion' and then progressively moves into positions in which it becomes more closely associated with the underlying metal structure, with a gradual loss of its water or ligand molecules. The final

94 Electrodeposition of metals

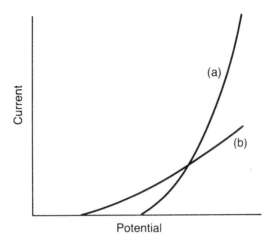

Figure 49. Simultaneous discharge of (a) zinc and (b) hydrogen

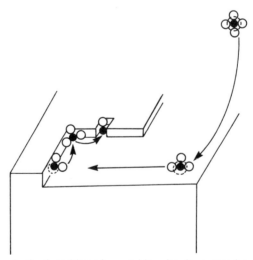

Figure 50. Stages in the deposition of a metal ion showing successive dehydration of the ion

position of the ion reflects the instability of an ad-atom on a plane surface. Kinks, steps and vacancies are preferred resting places for the deposited ion. The rate-determining step in the whole process is surface diffusion of the ad-ion at low overpotential and may be charge transfer at higher current densities and overpotentials. *See also* **Coulometry; Electroplating; Electrorefining; Electrowinning.**

Electrode potential (E; units: V). The potential of a half cell relative to that of a standard **hydrogen electrode** may be determined by combining it with a standard hydrogen electrode ($E^{\ominus}_{H+,H2} = 0$), preferably in a cell without a liquid junction, and measuring the e.m.f. of the cell. The e.m.f. of the cell, and hence the electrode potential of the half cell, depends on the activities of the oxidized and reduced forms in solution according to the general form of the **Nernst equation**. For the cell

$$\begin{array}{c|c|c|c} \text{Pt,H}_2(g) & \text{H}^+ & \text{oxidized and} & \text{Pt} \\ (101\,325 \text{ N m}^{-2}) & (a_{H+} = 1) & \text{reduced forms} & \end{array}$$

$$E_{cell} = E_{ox,red} - E_{H+,H2} = E_{ox,red}$$

$$= E^{\ominus}_{ox,red} + \frac{RT}{nF} \ln \frac{a_{ox}}{a_{red}} \qquad (58)$$

The standard electrode potential, $E^{\ominus}_{ox,red}$, is that potential which would be obtained when the oxidized and reduced forms are both present at unit activity. For cationic (metal) electrodes, the activity of the reduced form $a_M = 1$ and so Eq. (58) becomes

$$E_{Mz+,M} = E^{\ominus}_{Mz+,M} + \frac{RT}{zF} \ln a_{Mz+} \qquad (59)$$

The standard electrode (reduction) potential is positive (negative) when the electrode is more positive (negative) than the hydrogen electrode. If a cell is allowed to proceed spontaneously, the positive electrode is the one at which reduction (gain of electrons) occurs and the negative electrode is the one at which **oxidation** occurs.

All half cells (electrodes) can be arranged according to their standard electrode potentials (*see* Appendix, Table 3). If a cell is constructed with two metals dipping into their own solutions, with a bridge joining the two solutions, the metal which appears lower in the table is the positive pole and current flows inside the cell from the metal higher in the table (higher negative electrode potential) to that lower in the table. Thus metals with more negative electrode potentials tend to displace metals of more positive potential from solution.

An **equilibrium constant** and **solubility product** can be calculated from the standard electrode potentials.

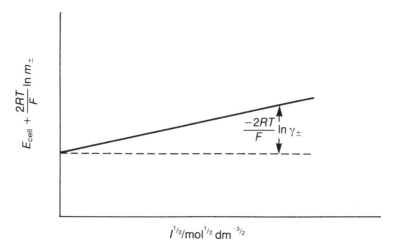

Figure 51. Graph to determine $E^{\ominus}_{cell}$

Determination of standard electrode potentials

1) From the variation, with the electrolyte concentration, of the e.m.f. of a simple cell containing a standard hydrogen electrode. For example

$$\ominus \quad \underset{(101\,325\text{ N m}^{-2})}{\text{Pt},\text{H}_2(g)} \;\Big|\; \underset{(m)}{\text{H}^+\text{Cl}^-} \;\Big|\; \text{AgCl},\text{Ag} \quad \oplus$$

for which

$$E_{cell} = E^{\ominus}_{cell} - \frac{2RT}{F}\ln a_\pm = E^{\ominus}_{cell} - \frac{2RT}{F}\ln m_\pm - \frac{2RT}{F}\ln \gamma_\pm$$

From the **Debye–Hückel activity equation**, $\log \gamma_\pm \propto I^{1/2}$; hence, the graph of $E_{cell} + (2RT/F) \ln m_\pm$ against $I^{1/2}$ is linear (*see* Figure 51). Since the mean ionic activity coefficient is unity at zero ionic strength the intercept is $E^{\ominus}_{cell}$, which is the standard electrode potential of the silver–silver chloride electrode (Eq. (58)). The deviation of the plot from the value for the standard electrode potential is a measure of the activity coefficient at various values of the molality.

Similar cells can be used for determining the standard e.m.f., and hence standard electrode potentials, in non-aqueous solvents.

2) From the variation, with electrolyte concentration, of the e.m.f. of galvanic cells which incorporate an electrode of the second kind (i.e. a reference electrode such as the **calomel electrode** or **silver–silver chloride electrode**). This method is particularly useful for metals that form highly dissociated chlorides. For example

$$\ominus \ \text{Zn} \ | \ \text{ZnCl}_2 \ | \ \text{AgCl}, \text{Ag} \ \oplus$$
$$(m)$$

for which

$$E_{cell} = E^{\ominus}_{cell} - \frac{RT}{2F} \ln 4m^3 - \frac{3RT}{2F} \ln \gamma_{\pm}$$

From this equation, the graph of $E_{cell} + (RT/2F) \ln 4m^3$ against $I^{1/2}$ is linear and of intercept $E^{\ominus}_{cell} = E^{\ominus}_{AgCl,Ag} - E^{\ominus}_{Zn2+,Zn}$, from which $E^{\ominus}_{Zn2+,Zn}$ can be obtained.

For bivalent metals that form highly dissociated soluble sulphates, similar cells using mercury–mercury (I) sulphate or lead–lead sulphate electrodes may be used. This method is widely used for the determination of electrode potentials in non-aqueous solutions; care must be exercised in the choice of electrodes to avoid solubility of the chloride or sulphate, or the formation of complexes with the solvent.

The method is also applicable to the determination of standard electrode potentials of the halogen electrodes, thus for the cell

$$\ominus \ \text{Ag,AgCl} \ | \ \text{MCl} \ | \ \text{Cl}_2(g), \text{Pt} \ \oplus$$

for which the reaction is

$$\text{Ag}(s) + \tfrac{1}{2}\text{Cl}_2(g) \rightarrow \text{AgCl}(s)$$

the e.m.f. is given by

$$E_{cell} = E^{\ominus}_{cell} + \frac{RT}{2F} \ln p_{Cl2}$$

The e.m.f. of the cell is independent of the nature or the concentration of the electrolyte, thus if E_{cell} is measured at known gas pressures the standard e.m.f. of the cell and hence the standard electrode potential of the chlorine electrode may be calculated.

3) Indirect method from equilibrium constants or tabulated standard free energy data. The method is applicable when the standard e.m.f. of the cell cannot be measured directly.

$$E^{\ominus}_{cell} = -\frac{\Delta G^{\ominus}}{nF} = \frac{RT}{nF} \ln K_{therm}$$

There are no new electrodes unique to non-aqueous solutions. Although some electrodes are not of much use in aqueous solutions, they may come into their own in non-aqueous solutions on account of possible changes in solubility relationships. Where metal salts dissolve and ionize in the solvent, the normal potential-determining processes occur at the metal electrode, giving rise to potentials defined by the Nernst equation

98 Electrode reaction mechanisms

$$E = E^{\ominus} + \frac{RT}{nF} \ln \alpha m \gamma$$

where α is the degree of ionization.

Electrodes of the second kind, based on sparingly soluble salts (e.g., of silver, mercury or lead) are of considerable use as they can be combined with other electrodes in cells without liquid junctions which can be used for the measurement of thermodynamic quantities.

Standard electrode potentials (all relative to the aqueous hydrogen electrode) in such solvents as methanol, acetonitrile, formamide, hydrazine and liquid ammonia (*see* Appendix, Table 4) have been obtained by measuring the e.m.f. of simple galvanic cells (*see* (2) above). See Covington and Dickinson (1973); Ives and Janz (1961).

Further reading
M. S. Antelman and F. J. Harris, *The Encyclopedia of Chemical Electrode Potentials*, Plenum Press, New York (1982)
R. G. Bates, Measurement of reversible electrode potentials, in *Techniques of Electrochemistry*, vol. 1, E, Yeager and A. J. Salkind (ed.) Wiley–Interscience, New York, p. 1 (1972)

Electrode reaction mechanisms. Mechanisms may be elucidated by studying current–voltage data under certain fixed conditions. The most simple is the measurement of steady-state current–voltage data when diffusion is not limiting, from which the **transfer coefficient** (α) may be determined. The transfer coefficient is a useful, although not unambiguous, parameter for the **hydrogen electrode reactions**. A more clear-cut example is given by the reaction at a copper electrode in an acidic copper (II) sulphate solution. Reduction or oxidation takes place via a two-step mechanism

$$Cu^{2+} + e \rightleftharpoons Cu^{+} \qquad (a)$$
$$Cu^{+} + e \rightleftharpoons Cu^{0} \qquad (b)$$

If the overpotential is sufficiently removed from equilibrium, the steady-state current density is related to the overpotential by the **Tafel equation**

$$i_c = i_{o,c} \exp\left(-\frac{\alpha_c F}{RT} \eta\right) \qquad (60)$$

$$i_a = i_{o,a} \exp\left(\frac{\alpha_a F}{RT} \eta\right) \qquad (61)$$

Considering the cathodic reaction, the two possible mechanisms have reaction (a) or reaction (b) as the rate-determining step. If reaction (a) is rate-determining, the overall current is given by

$$i_c = Fk_c a_{Cu2+} \exp\left(-\frac{\beta F}{RT}\eta\right) \qquad (62)$$

thus by comparison with Eq. (60), $\alpha_c = \beta \approx 0.5$. If reaction (b) is rate-determining, the current is

$$i_c = Fk_c a_{Cu+} \exp\left(-\frac{\beta F}{RT}\eta\right) \qquad (63)$$

Here the pre-exponential term includes the activity of the intermediate Cu^+ which is a function of overpotential. In this case, reaction (a) may be considered to be in pseudo-equilibrium. The forward and backward rates are equated

$$Fk_f a_{Cu2+} \exp\left(-\frac{\beta F}{RT}\eta\right) \approx Fk_b a_{Cu+} \exp\left[\frac{(1-\beta)F}{RT}\eta\right] \qquad (64)$$

Therefore

$$a_{Cu2+} = \frac{k_f}{k_b} a_{Cu2+} \exp\left(-\frac{F}{RT}\eta\right) \qquad (65)$$

As the overpotential becomes more negative, the concentration of Cu^+ increases. If the expression for a_{Cu+} is included in Eq. 63

$$i_c = Fk_c \frac{k_f}{k_b} a_{Cu2+} \exp\left[-\frac{(1+\beta)F}{RT}\eta\right] \qquad (66)$$

Comparison with Eq. 60 shows $\alpha_c = 1 + \beta \approx 1.5$. A similar process applied to the reverse reaction predicts $\alpha_a = \beta$ for a rate-determining step of

$$Cu^0 \rightarrow Cu^+ + e$$

and $\alpha_a = 2 - \beta$ for a rate-determining step

$$Cu^+ \rightarrow Cu^{2+} + e$$

Experimentally, $\alpha_c = 0.5$ and $\alpha_a = 1.5$.

General formulae may be derived which relate the cathodic and anodic transfer coefficients to the number of electrons required for the overall reaction (n), the stoichiometric number (ν), the number of steps preceding the rate-determining step (γ) and the symmetry coefficient of the rate-determining step (β).

$$\alpha_c = \frac{\gamma_c}{\nu} + \beta_c$$

$$\alpha_a = \frac{n - \gamma_c}{\nu} - \beta_c = \frac{\gamma_a}{\nu} + \beta_a$$

The stoichiometric number—the number of times the rate-determining step takes part in the overall reaction—may be used to distinguish between rival mechanisms in the **hydrogen electrode reactions**. The stoichiometric number is given by the sum of the anodic and cathodic transfer coefficients

$$\alpha_a + \alpha_c = \frac{n}{\nu}$$

or from the slope of the current–overpotential graph at low values of the overpotential

$$I = \frac{nI_0 F}{\nu RT}\eta$$

The above discussion assumes that the reaction is sufficiently well behaved to allow accurate measurement of current–voltage data over a wide range of overpotentials. For an irreversible system, such as the oxygen electrode at platinum, the linear region of the Butler–Volmer equation is not accessible. Double layer and diffusion effects may also make the linear region of the Tafel equation difficult to measure. Diffusion is best overcome by taking measurements before it becomes important using **transient methods** or by accurately controlling the diffusion layer in a **rotating-disc electrode** system.

In addition, the dependence of the current, at constant overpotential or voltage, on the concentration of reactant species may be measured. The pH dependence of the current is frequently used as a mechanistic indicator. The analysis of adsorbed intermediates, especially in organic electrochemical reactions, also gives a guide to the reaction mechanism. Fast reactions may be studied by **ac impedance** methods, although the analysis of the impedance data is somewhat complex. Spectroscopic techniques (*see* **Spectroelectrochemistry**), such as reflectance and surface Raman spectroscopy or **ellipsometry**, may give information concerning adsorbed intermediates, as may chronopotentiometry (*see* **Transient methods**) or **coulometry**. See Bockris and Reddy (1973).

Electrodialysis. An accelerated form of dialysis carried out in a three-compartment apparatus in which the central compartment is separated from the outer ones by a semipermeable membrane permeable to electrolyte ions. The colloid is contained in the central compartment and a potential is applied between platinum gauze electrodes mounted in the surrounding solvent or dilute electrolyte solution. The ions to be removed move in opposite directions and may be washed away with fresh solvent.

Industrial electrodialysis units consist of alternating series of cation- and anion-permeable membranes placed between two electrodes (*see* Figure 52). Feed water flows through the stack parallel to the membranes and the current

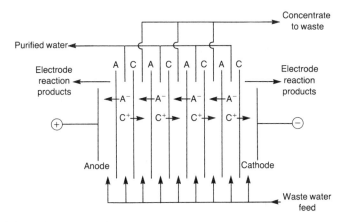

Figure 52. Schematic representation of industrial electrodialysis unit: A, anion-permeable membrane, A^- anion; C, cation-permeable membrane, C^+ cation

is passed normal to the plane of the membranes. As a result of selective transport of anions and cations across membranes, alternate cells are depleted of solute, whereas cells adjacent are concentrated in solute. The depleted solution is taken off as purified water, and the concentrate and electrode reaction products flow to waste.

For efficient economic operation, the membranes should have the following properties: (1) high degree of selectivity for ions of a given sign; (2) low electrical resistance, with high water content and high concentration of fixed charges; (3) low degree of water transport (electroosmosis reduces efficiency and must be avoided); (4) inertness with no deterioration in the presence of chemical and biological material; (5) physical strength; (6) resistance to deterioration at high temperatures.

Electrodialysis is widely used in the desalination of brackish water, especially slightly brackish water where there is a relatively small amount of contaminant to be removed. It is preferred over other methods (e.g., reverse osmosis, freezing and flash distillation) on economic grounds. The optimum operational temperature for electrodialysis plants desalinating sea water is about 70°C. The method is of use in sewage works and in the recycling of water to prevent the build-up of mineral salts which occurs with the usual methods of treatment.

Further reading
I. F. Miller, Electrodialysis of aqueous solutions, in *Techniques of Electrochemistry*, vol. 3, E. Yeager and A. J. Salkind (ed.), Wiley–Interscience, New York, p. 437 (1978).

102 Electroflotation

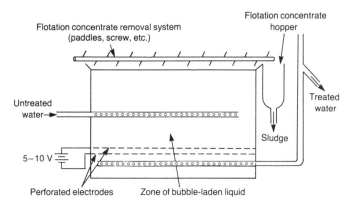

Figure 53. Diagrammatic representation of electroflotation plant

Electroflotation. A method of separating dilute suspensions into slurries and clear liquid. Such suspensions can be treated merely by allowing them to stand in settling tanks or by the addition of flocculating agents: both of which are very costly. When electrolytically generated bubbles (which are very small with highly favourable surface area/volume ratios) are allowed to rise up through a suspension or colloidal liquid, a speedy separation is achieved. The process is believed to be mechanical, involving the lifting of the suspended particles by the minute bubbles generated. There may, however, be an additional effect relating to the charge carried by the bubbles, colloidal particles, etc.

The solid or oily matter which rises to the top of the cell (*see* Figure 53) is removed mechanically by a bladed conveyor. The electrodes used can be platinized titanium, carbon steel, stainless steel or lead dioxide. Electroflotation is employed in such processes as the treatment of steel rolling mill wastes, paper mill effluents, scouring liquors from yarn and for dealing with abattoir wastes.

Electroforming. A process in which a thick deposit of metal is electroplated on to a mould, which may be of metal or other material (such as plastic), which has been given a conducting coating. The deposit is then removed and gives a perfect replica of the surface plated. The process is therefore a form of **electroplating** and involves no new principles. Uses include the making of electrotype by plating copper on to a plastic sheet bearing the required impression, and a somewhat similar process is used for making master plates from which gramophone records are pressed.

Electrogravimetric analysis. *See* **Coulometry.**

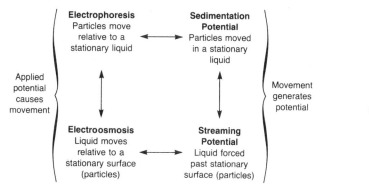

Figure 54.

Electrokinesis. A group of phenomena which have a common origin in the asymmetrical distribution of ions at an interface, the **electrical double layer**. At an interface between two phases, on the time average, ions of one kind will be associated with one phase and those of the opposite sign with the other phase. When an electric field is applied across such a system, there will be a tangential movement of one phase relative to the other at a velocity that depends on the potential at the plane of shear, the electrokinetic or ζ-**potential**. Conversely an induced relative movement of the two phases will give rise to a measurable potential difference, the value of which is also dependent on the ζ-potential.

The four electrokinetic effects are related (*see* Figure 54).

Studies of these phenomena provide valuable information about the structure of the double layer and information about the separation and identification of components in mixtures.

Electrokinetic phenomena. *See* **Electrokinesis.**

Electrokinetic potential. *See* ζ-**potential.**

Electrolysis of brine. Sodium chloride solutions are electrolysed on a very large scale for the production of sodium hydroxide and chlorine. Electrolysis is also used to produce hypochlorites (ClO^-) and chlorates (ClO_3^-).

The potential required to deposit sodium from an aqueous solution is -2.71 V, so that at a solid cathode hydrogen is always liberated in preference; 1 mole of hydroxide ion per faraday is formed at the same time. If mercury is used as the cathode, the discharge of sodium ions will not produce a surface film of sodium metal, so the value -2.71 V is irrelevant. The sodium amalgamates with the mercury and diffuses away into the interior, so that the activity of sodium at the surface is extremely small. This, combined with the

104 Electrolysis of brine

very high hydrogen overpotential (*see* **Hydrogen electrode reactions; Overpotential**) at a mercury cathode, especially at fairly high current densities, ensures that sodium discharge is the predominant reaction. At the anode, the primary reaction is the discharge of chloride ions to give chlorine gas (*see* **Halogen electrode reaction**). Here, again, thermodynamic considerations would suggest that oxygen from the water ($E^\ominus = +0.82$ V at pH 7) would be discharged more readily than chlorine ($E^\ominus = +1.36$ V). It is not, partly because oxygen discharge would leave a highly acid film in contact with the anode, which would raise the theoretical E by several tenths of a volt, and partly because of the high oxygen overpotential.

If the primary products are allowed to mix, hypochlorite is formed by the reaction

$$Cl_2 + 2OH^- \rightarrow Cl^- + ClO^- + H_2O$$

and if all the liberated chlorine is combined in this way, 2 faradays of electricity produce 1 mole of hypochlorite. The hypochlorite may react further in two ways. In a warm acid solution, chlorate is formed by the reaction

$$2HClO + ClO^- \rightarrow ClO_3^- + 2H^+ + 2Cl^-$$

1 mole of chlorate resulting from the passage of 6 faradays. This reaction is very slow at low temperatures and in alkaline solutions, and the hypochlorite then undergoes a further reaction at the anode

$$12ClO^- + 6H_2O \rightarrow 4ClO_3^- + 12H^+ + 8Cl^- + 3O_2 + 12e$$

Here 36 faradays have produced 4 moles of chlorate (and three of oxygen), so that chlorate production is less efficient by this route. These considerations form the basis for the following processes:

Sodium hydroxide and chlorine production.
1) Diaphragm cells. A steel cathode and **dimensionally stable anode** are used. Formerly carbon was the anode material. Although inexpensive, these anodes have a high corrosion rate which has led to dimensionally stable anodes superseding graphite anodes. Hydroxide ions can reach the cathode by diffusion or by electrolytic migration. To prevent the former, a diaphragm of treated asbestos surrounds the cathodes. Electrolytic transport is mainly by the chloride ions, which are present at a high concentration, but hydroxide ions have a very high mobility, and to prevent them reaching the anode it is necessary to use countercurrent circulation. The brine is fed continuously into the anode compartments, from which the chlorine is collected, and flows out through the cathode compartments; it is then concentrated, and the remaining sodium chloride crystallized out. Hydrogen from the cathode compartments is also collected.

2) **Membrane cells.** A disadvantage of an asbestos diaphragm is that it is not possible to maintain a high hydroxide concentration in the cathode compartment. Thus considerable work-up is required to concentrate the 10 percent hydroxide solution which flows out of the cell. Membranes based on polyfluorosulphonic acids and polyfluorocarboxylic acids do have the necessary selectivity, allowing concentrations of hydroxide up to 30 percent (*see* **Solid electrolytes**).

3) **Mercury cells.** These are usually of (nearly) horizontal construction in which the mercury forming the cathode flows slowly along the floor of the cell in countercurrent to the brine solution. Graphite anodes dip into the latter, and the chlorine liberated is led off. The amalgam concentration is not allowed to reach 0.2 percent sodium, as otherwise the fluidity of the mercury decreases, and also an appreciable amount of hydrogen would be discharged which would lower the current efficiency. The amalgam flows from the cell to a decomposer where it reacts with water

$$Na(Hg) + H_2O \rightarrow Hg + NaOH + \tfrac{1}{2}H_2$$

Pure concentrated sodium hydroxide is produced directly by this method, but the voltage required is higher than in a diaphragm or membrane cell. Theoretically some of the extra electrical energy could be recovered from the spontaneous reaction in the decomposer, but practical difficulties have so far prevented this. The alkali metal amalgams produced in mercury cells also have uses as reagents for various organic syntheses.

Hypochlorite production. Sodium chloride is electrolysed between electrodes of graphite. The temperature is kept low to minimize chlorate formation, and the electrodes are kept close together and arranged to ensure mixing. However, any hypochlorite reaching the cathode will be reduced, and in large plants it is more common to produce sodium hydroxide and chlorine in diaphragm or membrane cells and form hypochlorite by mixing these.

Where sea water is available, electrolysis provides an efficient method of sewage treatment. The sewage is mixed with sea water and passed through an electrolytic cell. Hypochlorite is formed and this rapidly sterilizes the sewage which then sediments out; the latter process is assisted by the gelatinous precipitate of magnesium hydroxide which forms in the alkaline solution.

Chlorate production. Sodium chloride is electrolysed between steel cathodes and dimensionally stable anodes. The temperature is kept at about 50°C and a pH of 6 is maintained. A concentrated solution of sodium chlorate is produced which crystallizes on cooling.

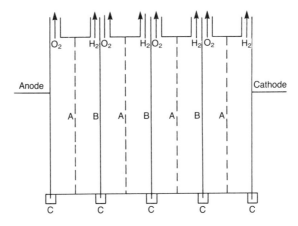

Figure 55. Electrolysis of water in bipolar cell: A, diaphragms; B, bipolar electrodes; C, insulators

Electrolysis of water. Hydrogen and oxygen can be produced by a variety of methods, but where electricity is cheap the electrolysis of water has the advantage of producing gases of high purity. An electrolyte must be present to give high conductivity; the high mobilities of the proton and hydroxyl ion suggest the use of acids or alkali, but common materials withstand attack by alkalis much better than by acids, so sodium and potassium hydroxides at a concentration giving the maximum conductance are commonly used. Distilled water is added continuously to maintain these conditions, and the operating temperature is about 343 K. The decomposition voltage of water is 1.23 V, but the **overpotential** and resistance losses bring up the working voltage to about 2 V. The majority of the overpotential of the cell comes from the oxygen electrode (*see* **Oxygen electrode reactions**). Recently, transition metal oxide anodes of the spinel structure ($NiCo_2O_4$ and lithium-doped Co_3O_4) have been shown to reduce the operating voltage to 1.8 V. The current efficiency can be as high as 99 percent, and the hydrogen produced is about 99.9 percent pure after drying.

The cathode is normally of steel and is separated from the anode by an asbestos membrane. A bipolar arrangement is often adopted (*see* **Bipolar electrode**), as shown diagrammatically in Figure 55.

Heavy water production. When water is electrolysed, the lightest isotope of hydrogen is preferentially liberated owing to differences in the activation energy for the reactions at the cathode. There is therefore a gradual enrichment in deuterium of the liquid phase. A practical measure of the extent of the effect is given by the separation factor S where

$$S = \frac{(H/D)_{gas}}{(H/D)_{liquid}}$$

Under the conditions used, the value of S is about 6. A number of cells are arranged in a cascade so that the deuterium-enriched product of one cell feeds the next. Almost pure heavy water (D_2O) can be prepared in this way.

Electrolytic cell. The interconversion of electrical and chemical energy can be effected in a primary cell, in which a spontaneous chemical process is used to produce electricity; or in the converse process of electrolysis, in which electrical energy from an outside source is used to bring about a chemical reaction.

An electrolytic cell normally consists of two electrodes dipping into an electrolyte solution or of two solutions in contact through a porous diaphragm, with a suitable electrode in each. The electrode material will be chosen for its chemical properties and also, perhaps, for its catalytic effect upon the desired reaction at its surface. When such a cell is set up, equilibrium conditions are rapidly established; the electrodes acquire potentials at which there is no net tendency for electrons to move in either direction across the electrode/solution interface, and the corresponding **electrical double layer** is set up at each electrode surface. If now a gradually increasing e.m.f. is applied from an external source, electrolysis will begin when this exceeds the equilibrium voltage. The availability of electrons at one electrode (the cathode) is now high enough to cause a reduction process (e.g., $e + Ag^+ \rightarrow Ag$ or $e + Fe^{3+} \rightarrow Fe^{2+}$) to proceed at a finite rate, while at the other electrode (anode) an oxidation process (e.g., $I^- \rightarrow \frac{1}{2}I_2 + e$ or $Ag \rightarrow Ag^+ + e$) supplies electrons to the electrode. The external battery can be likened to an electron pump, transferring electrons from anode to cathode. The corresponding current passing through the solution is carried by ions migrating to the electrodes.

A part of the electrical energy is expended in overcoming the resistance of the electrolyte. From Ohm's law, this voltage drop is given by $\Delta E = IR$ for a current I A passing through a resistance of R Ω. The remainder of the applied e.m.f. must exceed $E^\ominus$, the equilibrium value of the cell reaction, for electrolysis to occur; $E - E^\ominus$, the driving force of the reaction, is called the **overpotential**. The rate of the reaction is measured by the magnitude of the current passing, and this increases (within limits) with increasing overpotential. The extent of chemical reactions is governed by **Faraday's laws**.

Examples.
1) The current-carrying ions are discharged at the electrodes. In the electrolysis of moderately concentrated hydrogen chloride solution, with inert platinum electrodes and a voltage of about 1.3 V, hydrogen is evolved at the

cathode and chlorine at the anode. In the solution, the current is carried by the migration of hydrogen ions to the cathode and chloride ions to the anode; in the outside circuit electrons flow from anode to cathode

$$H_3O^+ + e \rightarrow \tfrac{1}{2}H_2 + H_2O \quad \text{at cathode}$$
$$Cl^- \rightarrow \tfrac{1}{2}Cl_2 + e \quad \text{at anode}$$

2) A difficultly discharged anion leads to the decomposition of water at the anode to give hydrogen ions, oxygen and electrons (e.g., electrolysis of copper sulphate solution with inert platinum electrodes. At the cathode, copper is deposited: $Cu^{2+} + 2e \rightarrow Cu$. At the anode, oxygen is evolved and the solution adjacent to the electrode contains sulphuric acid. Sulphate ions migrate to the anode but are not the species discharged. Thus the anode reaction is

$$2H_2O \rightleftharpoons 2H^+ + 2OH^-$$
$$2OH^- \rightarrow H_2O + \tfrac{1}{2}O_2 + 2e$$

Overall $\quad\quad\quad\quad 2H_2O \rightarrow 2H^+ + \tfrac{1}{2}O_2 + 2e$

3) A difficultly discharged cation leads to the decomposition of water at the cathode with the production of hydroxide ions and hydrogen (e.g., electrolysis of sodium chloride solution with platinum electrodes). The cathode reaction is represented by

$$H_2O \rightleftharpoons H^+ + OH^-$$
$$H^+ + e \rightarrow \tfrac{1}{2}H_2$$

overall $\quad\quad\quad\quad H_2O + e \rightarrow OH^- + \tfrac{1}{2}H_2$

4) Reacting electrodes (e.g., the electrolysis of copper sulphate solution between copper electrodes). Copper (II) and sulphate ions are the current-carrying ions. Copper is deposited on the cathode and copper passes from the anode into solution, there being no change in the electrolyte concentration. Cells of this type are used in the **electrorefining** of metals and as a **coulometer**.

Electrolytic oxidation of organic compounds. See **Organic electrochemistry; Organic synthesis.**

Electrolytic polishing. Selective dissolution of metals in which projections in the surface are attacked more rapidly than depressions. The object may be to produce a bright surface or one suitable for microgravimetric examination. The surface to be treated is made the anode, and a great variety of electrolytes have been used; perchloric, chromic and nitric acids are common constituents, but salt solutions or fused salts are more suitable in some cases.

The technique has not been fully explained, but the conditions under which electropolishing is most effective appear to be those in which the anodic dissolution of the metal is diffusion controlled. As the metal dissolves, there is a layer of solution of high concentration at the anode, and this may give rise to a film of insoluble salt or oxide at the surface which hinders the diffusion of ions into the bulk of the solution. These factors reinforce the normal tendency for diffusion to take place more readily from projections than from concave areas of the surface, and thus give rise to the polishing effect.

Electrolytic reduction of organic compounds. *See* **Organic electrochemistry; Organic synthesis.**

Electrometric titrations. In all electrometric methods of titration, the precise location of the 'end-point' is obtained from a series of independent observations, rather than from one estimate of the end-point as in the indicator method. To overcome some of the limitations of visual indicators (e.g., in highly coloured solutions) and to extend the advantages of titrimetric analysis to titrations not otherwise feasible because of adverse equilibrium constants for the reaction, instrumental methods of analysis have been developed to locate the end-point.

In potentiometric titrations, the change in the e.m.f. of a suitable cell (containing an indicating and a reference electrode) is dependent on the logarithm of the concentration of the species being titrated, and so it is important to obtain a large number of readings in the vicinity of the equivalence point. In amperometric and conductiometric titrations, the equivalence point is determined by extrapolation from readings taken remote from the equivalence point. Conductiometric are preferred to potentiometric titrations in the titration of weak acids and weak bases and for titrations at very low concentrations.

Amperometric titrations. The polarographic determination of trace amounts of electroactive substances can be used to extend the range of volumetric titrations to lower concentrations by sensing the end-point of a titration from the abrupt change in the limiting current on the addition of a titrant. A complete polarogram (*see* **Polarography**) after each addition of titrant is not required; it is only necessary to apply a sufficiently large, but constant, potential between the dropping mercury electrode or rotating platinum electrode so that the measured current is strictly the limiting current.

Figure 56(a) illustrates the addition of sulphuric acid to a solution of lead nitrate in nitric acid. As the acid is added the electroactive lead ions are removed by precipitation, and this results in a reduction of the limiting current, which falls to a low value at the end-point and remains at this value even on the addition of more sulphuric acid. The reduction of the hydrogen ion does not occur at the minimum potential at which the limiting diffusion

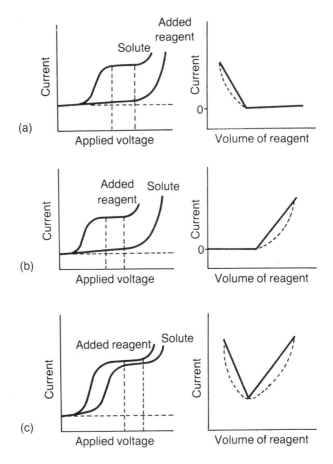

Figure 56. Typical amperometric titration curves: (a) titration of lead salt with sulphuric acid; (b) titration of a sulphate with a barium salt; (c) titration of a lead salt with a dichromate solution

current of lead is established. There is a degree of curvature especially near the end-point, which can be accurately determined by extrapolation of the linear portions of the graph. Allowance for the volume of reagent added must be made.

When the added reagent gives a diffusion current while the material being titrated does not, the diffusion current remains at zero until the end-point is reached; after which it increases (e.g., the titration of a sulphate with a solution of a barium salt, see Figure 56(b)). If both substances are electroactive, a V-shaped titration curve (see Figure 56(c)) arises, one arm representing the diffusion current of the reagent and the other that of the solute (e.g., the titration of a lead salt with a solution of a dichromate).

Electrometric titrations

Conductiometric titrations. The addition of one electrolyte (A^+B^-) to another (C^+D^-) will result in a change of conductance on account of volume changes and possible ionic reactions. If the addition is made so that there is no appreciable change in volume and there is no chemical reaction, the conductance of CD will gradually increase on the addition of AB. On the other hand, if there is an ionic reaction

$$A^+B^- + C^+D^- \rightarrow A^+D^- + CB$$

in which one of the products (CB) is either only slightly ionized or insoluble, then a marked change of conductance occurs at the equivalence point. During the addition, the ions A^+ replace the ions C^+ and the conductance may increase or decrease depending on the relative mobility values of A^+ and C^+. Figures 57(a), (b) and (c) illustrate typical conductiometric titration curves. The more acute the angle at which the lines intersect the greater is the accuracy of locating the end-point. A volume correction is necessary unless the concentration of the titrant is about 100 times greater than that of the reagent being titrated.

High-frequency conductiometric titrations have certain advantages over the conventional technique in that the electrodes are not in contact with the solution (thereby eliminating errors due to electrode contamination) and titrations in non-aqueous solvents can be followed more readily and mixtures analysed.

Potentiometric titrations. An electrochemical cell can be constructed using an indicator electrode coupled with a reference electrode so that the **electrode potential** of the indicating electrode, and hence the e.m.f. of the cell, is a measure of the activity (concentration) of the ionic species in solution. Thus for the titration of an acid with a base using the cell

$$\ominus \; Pt, H_2(g) \,|\, H^+ \;\vdots\; KCl \,|\, Hg_2Cl_2, Hg \; \oplus$$

$$E_{cell} = E_{calomel} - E_{H^+,H_2} = E' - \frac{RT}{F} \ln c_{H^+}$$

$$= E' + 0.0591 \, pH$$

On the addition of the base, the e.m.f. or pH changes relatively slowly at first since this depends on the fraction of the particular ion removed. Towards the equivalence point the fraction of ion (H^+) removed by a constant amount of titrant (base) increases rapidly, and this is reflected by a rapid change in the e.m.f. (pH). Above the equivalence point, the curve flattens out (*see* Figure 58). The exact location of the equivalence point is best achieved by plotting the first derivative curve (i.e. $\Delta pH/\Delta V$ or $\Delta E/\Delta V$ against V).

The following conditions must be fulfilled by a reaction for it to be used in a potentiometric titration: (1) there must be a fixed stoichiometric relationship

112 Electrometric titrations

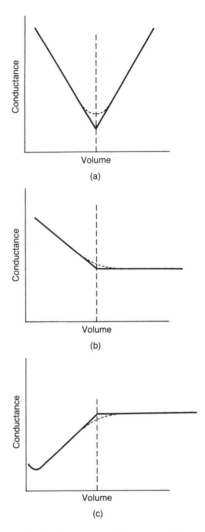

Figure 57. Typical conductiometric titration curves: (a) strong acid–strong base, precipitation titration; (b) strong acid–weak base; (c) weak acid–weak base

between the two reagents; (2) an indicator electrode must be available which has an electrode potential determined uniquely and reversibly by the concentration of one ion in the reaction; (3) the reaction must be as complete as possible; and (4) rapid equilibrium must be established between the reacting species and the indicator electrode.

1) Acid–base titrations. The shape and position of the titration curve (*see*

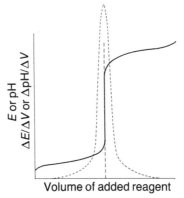

Figure 58. Typical potentiometric titration curve and first derivative plot (-----)

Figure 59) obtained using a hydrogen-indicating electrode in combination with a reference electrode: for example

$$\text{glass electrode} \mid H^+ \mid KCl \mid Hg_2Cl_2, Hg$$

depends on the strengths of the acid and the base. The curve for the titration of a strong acid with a strong base (I–II) is readily calculated since the concentration of hydrogen ions is equal to the concentration of unneutralized acid up to the equivalence point (B), at which point the concentration is 10^{-7} mol dm^{-3}.

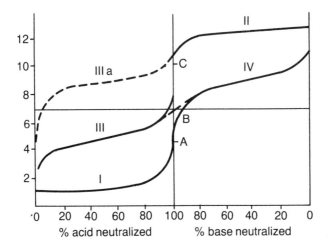

Figure 59. Neutralization curves for acids and bases (all concentrations 0.1 mol dm^{-3}): I, strong acid; II, strong base; III, weak acid; IIIa, very weak acid; IV, weak base. A is equivalence point for I–IV, B for I–II and C for II–III

114 Electrometric titrations

The curve for the titration of a weak acid with a strong base (III–II) can be calculated up to the equivalence point (C) by the Henderson equation

$$\mathrm{pH} = \mathrm{p}K_a + \log \frac{c_{A^-}}{c_{HA}}$$

At the end-point, the solution is alkaline owing to the hydrolysis of the salt of the weak acid and strong base; the pH is given by

$$\mathrm{pH} = \tfrac{1}{2}\mathrm{p}K_w + \tfrac{1}{2}\mathrm{p}K_a + \tfrac{1}{2}\log c$$

Similarly for the titration of a weak base with a strong acid (I–IV), the pH at the equivalence point (A) on the acid side of neutrality is given by

$$\mathrm{pH} = \tfrac{1}{2}\mathrm{p}K_w - \tfrac{1}{2}\mathrm{p}K_b - \tfrac{1}{2}\log c$$

In the titration curve for a weak acid–weak base (III–IV), there is only a gradual change of pH throughout the titration with no marked change at the equivalence point.

2) **Precipitation titrations.** These are generally limited to those using silver nitrate as one reagent with a silver indicating electrode and a reference electrode separated from the solution by a **salt bridge**

$$\ominus \ \mathrm{Hg, Hg_2Cl_2} \,|\, \mathrm{KCl} \,\vdots\, \mathrm{NH_4NO_3} \,\vdots\, \mathrm{halide\ solution} \,|\, \mathrm{Ag} \ \oplus$$

for which

$$E_{\mathrm{cell}} = E_{\mathrm{Ag^+,Ag}} - E_{\mathrm{cal}} = E^{\ominus}_{\mathrm{Ag^+,Ag}} - E_{\mathrm{cal}} + \frac{RT}{F}\ln a_{\mathrm{Ag^+}}$$

On the addition of the first drop of silver nitrate solution silver chloride is formed, and the solution becomes saturated with respect to it; the concentration of silver ions is thus small and is indicated by the potential of the silver electrode. As more silver nitrate is added more silver chloride is precipitated, and the solution remains saturated with silver chloride but the concentration of the silver ion increases slightly to keep the solubility product of silver chloride constant as the chloride ion is removed. At the end-point, the concentration of silver ions increases rapidly owing to the presence of excess silver ions and the electrode potential of the silver electrode increases rapidly (*see* Fig. 58).

With such a cell it is possible in a single titration to determine the concentrations of iodide, bromide and chloride ions in solution. The different solubility products are reflected in three distinct steps in the e.m.f. of the cell corresponding to the complete precipitation of silver iodide, silver bromide and silver chloride.

3) **Redox titrations.** The position of the first part of the curve (*see* Figure 60) is determined by the standard electrode potential of the titrated system and

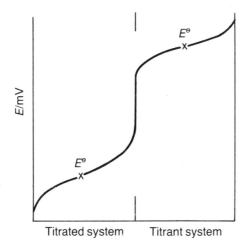

Figure 60. Oxidation–reduction titration curve

the second half by the standard electrode potential of the titrant. For the general case of two redox systems involving different numbers of electrons

$$\text{red}_2 \rightarrow \text{ox}_2 + ae$$
$$\text{ox}_1 + be \rightarrow \text{red}_1$$

the cell reaction is

$$a\,\text{ox}_1 + b\,\text{red}_2 \rightleftharpoons a\,\text{red}_1 + b\,\text{ox}_2$$

The potential at the equivalence point (E_e) which is equal to the electrode potential of each of the separate systems is given by

$$E_e = \frac{aE_2^\ominus - bE_1^\ominus}{a+b}$$

The equilibrium constant can be calculated using the equation

$$\ln K = nF\frac{(E_2^\ominus - E_1^\ominus)}{RT}$$

In the titration of an iron (II) salt with potassium permanganate using the cell

$$\ominus\ \text{Pt}\,|\,\text{Fe}^{2+},\text{Fe}^{3+}\,\vdots\,\text{KCl}\,|\,\text{Hg}_2\text{Cl}_2,\text{Hg}\ \oplus$$

the reaction is

$$5\text{Fe}^{2+} + \text{MnO}_4^- + 8\text{H}^+ \rightarrow 5\text{Fe}^{3+} + \text{Mn}^{2+} + 4\text{H}_2\text{O}$$

for which $E_e = 1.39$ V and $\log K = 63.5$. At the equivalence point $c_{\text{Fe}^{3+}}/c_{\text{Fe}^{2+}} = 4 \times 10^{10}$. This large value of the equilibrium constant is brought about by the large difference in the standard electrode potentials of the two systems. If a given reaction is to be suitable for volumetric analysis, it is necessary that

116 Electromotive force

$K > 10^6$. This corresponds to a minimum difference in the standard electrode potentials of 0.35 V if $n = 1$ for both systems, 0.26 V if $n = 1$ and $n = 2$ for the two systems and 0.18 V if $n = 2$ for both systems. *See also* **Redox electrode systems.** See James and Prichard (1977).

Further reading
D. A. Skoog and D. M. West, *Fundamentals of Analytical Chemistry*, Holt-Saunder International, New York (1976)
D. A. Skoog and D. M. West, *Principles of Instrumental Analysis*, Saunders College, New York (1980)

Electromotive force (e.m.f.) (dimensions: $\varepsilon^{-1/2} m^{1/2} l^{1/2} t^{-1}$; units: V = kg m² s⁻³ A⁻¹ = J A⁻¹ s⁻¹). That force which causes a current to flow between two points. The unit of electric potential is the **volt**. The sign of the electromotive force (e.m.f.) is defined so that a positive charge will tend to flow from a higher to a lower potential. Applied to a **cell**, the e.m.f. is the electric potential between two pieces of metal with identical composition, the ends of the chain of conducting phases. Digital voltmeters have replaced potentiometers for the measurement of e.m.f. values.

Electron transport chain. In biological systems oxygen is the final oxidizing agent, but rarely is there a direct reaction in which the reduced form of the metabolite (MH_2) reacts with oxygen as would be represented by

$$MH_2 + \tfrac{1}{2}O_2 \rightarrow M + H_2O$$

rather is there a transfer of electrons from the substrate to a coenzyme

$$MH_2 + CoE \rightarrow M + CoEH_2$$

with the subsequent oxidation of the reduced coenzyme. The hydrogen atoms on the reduced coenzyme are removed by a second enzyme linked by sharing of the same cofactor; this leads to electron transfer in a sequence running from MH_2 to O_2.

Coenzymes involved in the electron transfer processes are highly specific, the most common ones are nicotinamide adenine dinucleotide (NAD^+) and the corresponding phosphate ($NADP^+$). The large ribose and adenine structures in NAD^+ and $NADP^+$ are required for specificity, the nicotinamide is the moiety responsible for the electron transfer process

or

$$NAD^+ + 2H^+ + 2e \rightleftharpoons NADH_2^+ \text{ or } NADH + H^+$$

Flavin adenine dinucleotide (FAD), coenzyme Q_{10} and the cytochromes are other examples of cofactors in the transport chain.

During the anaerobic breakdown of glucose by animal muscle, glyceraldehyde-3-phosphate, in the presence of inorganic phosphate, is oxidized to 1,3-diphosphoglyceric acid (catalysed by glyceraldehyde-3-phosphate dehydrogenase); the hydrogen atoms are transferred via $NAD^+/NADH_2^+$ to pyruvic acid which is reduced to lactic acid (catalysed by lactate dehydrogenase)

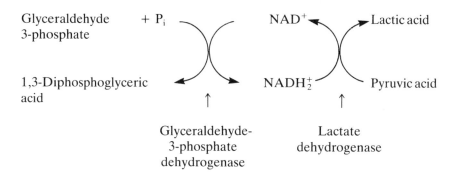

Aerobically the degradation of glucose proceeds through: (1) the Kreb's (or tricarboxylic acid) cycle; and (2) the sequential electron transport chain, in which oxygen is the terminal oxidizing agent, and carbon dioxide and water are the end-products. Pyruvic acid, which enters the Kreb's cycle via acetyl coenzyme A, is converted into 3 mole of carbon dioxide; at the same time, the reduced forms of coenzymes ($NADH_2^+$ and $FADH_2^+$) are formed. These molecules are oxidized, and the electrons (but not hydrogen atoms) are transferred down a series of cytochromes (proteins containing a haem group) which serve as redox catalysts in cell respiration. The terminal carrier of the electron transport chain is the cytochrome a–cytochrome a_3 complex, sometimes called cytochrome oxidase. This transfers the electrons directly to a molecule of oxygen which is reduced to water. The sequence of the carriers in the electron transport is determined by the relative electrode potentials of the **redox electrode systems**. Table 12 represents the sequential electron transport in mitochondria; each step involves the transfer of two electrons. Values of $E^{\ominus\prime}/V$ for each redox system and $\Delta G^{\ominus\prime}/kJ\,mol^{-1}$ for each reaction are given.

Some of the free energy liberated in the electron transport process is involved in the synthesis of 3 mole of adenosine triphosphate (ATP) from adenosine diphosphate (ADP) and inorganic phosphate at three different sites by a process known as oxidative phosphorylation. The energy is stored in ATP until required for coupling with endergonic processes in cell metabolism

Electron transport chain

Table 12
Schematic representation of the electron transport chain in mitochondria

Reaction	$E^{\ominus\prime}/V$	$\Delta G^{\ominus\prime}/\text{kJ mol}^{-1}$
Kreb's cycle ↑↓ NAD⁺ / NADH$_2^+$ → ADP + P$_i$ → ATP	−0.32	−40.5
FADH$_2$ / FAD ⇌ CoQ / CoQH$_2$	−0.11	−40.5
	0.10	+9.65
cyt b.Fe^{2+} / cyt b.Fe^{3+} → ADP + P$_i$ → ATP	0.05	−40.5
cyt c.Fe^{3+} / cyt c.Fe^{2+}	0.26	−9.65
cyt a.Fe^{2+} / cyt a.Fe^{3+} → ADP + P$_i$ → ATP	0.29	−46.3
cyt a_3.Fe^{3+} / cyt a_3.Fe^{2+}	0.53	−55.97
OH⁻ + H⁺ ↓ H$_2$O / ½O$_2$	0.82	

The overall reaction is

$$\text{NADH}_2^+ + 3\text{ADP} + 3\text{P}_i + \tfrac{1}{2}\text{O}_2 \rightarrow \text{NAD}^+ + 3\text{ATP} + \text{H}_2\text{O}$$

for which the change in electrode potential is 1.14 V and the standard free energy change $\Delta G^{\ominus\prime} = -220.0 \text{ kJ mol}^{-1}$

when it is released by hydrolysis: for example in the phosphorylation of glucose

$$\text{ATP} + \text{H}_2\text{O} \rightarrow \text{ADP} + \text{H}^+ + \text{P}_i \qquad \Delta G^{\ominus\prime} = -29.3 \text{ kJ}$$
$$\text{Glucose} + \text{P}_i \rightarrow \text{Glucose-6-P} + \text{H}_2\text{O} \qquad \Delta G^{\ominus\prime} = +12.5 \text{ kJ}$$

Thus

$$\text{ATP} + \text{Glucose} \rightarrow \text{Glucose-6-P} + \text{ADP} \qquad \Delta G^{\ominus\prime} = -16.8 \text{ kJ}$$

The remainder of the free energy released in small amounts along the electron transport chain is dissipated as waste energy. *See also* **Biological oxidation–reduction systems;** Appendix, Table 6.

Electroosmosis. One of the electrokinetic phenomena dependent on the existence of an **electrical double layer** at an interface. If a potential difference is applied between the ends of a capillary tube containing electrolyte, or across a plug of finely divided material (which can be regarded as a bundle of capillaries), a movement of the liquid is observed. This is the reverse of **electrophoresis**, where particles under an applied field move through a stationary liquid. The effect can be studied in an apparatus such as that shown in Figure 61. A plug of the finely divided material is in the centre of the tube, which is completely filled with electrolyte solution, and a potential of about 200 V is applied between the two electrodes (calomel or silver–silver chloride). For steady-state conditions, the electrical force applied must balance the frictional force.

For a plug of material of length l and effective cross-sectional area A, the volume of liquid transported in unit time (measured by the rate of movement of the air bubble) is

$$V = \frac{A\zeta\varepsilon E}{4\pi\eta l}$$

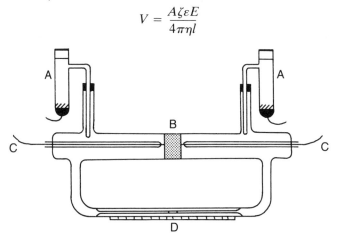

Figure 61. Electroosmosis cell: A. calomel electrodes; B, plug of material; C, leads to platinum gauze electrodes; D, air bubble and scale

The area cannot be measured accurately, but if I, the current passing, is measured and the conductivity κ of the electrolyte is known (determined from the conductance between the two platinum gauze electrodes which also hold the plug in place), then

$$I = \frac{E\kappa A}{l}$$

Thus

$$V = \frac{\zeta \varepsilon I}{4\pi \eta \kappa}$$

Values of the ζ-**potential** obtained from electroosmotic studies for a given solid/electrolyte interface are in close agreement with those obtained from electrophoresis and sedimentation potentials. Since most interfaces of interest exist as particulate matter (colloids, biological cells) it is more convenient to use electrophoretic techniques to determine their ζ-potentials, rather than to compress them into a plug.

Electroosmotic streaming of liquid on surfaces and through the pores of filter paper cannot be eliminated and must be taken into account in determining the electrophoretic mobility.

Electrophoresis. Migration of ions, colloidal particles and particulate matter through a liquid under an applied electrical potential. The separation and identification of components in a mixture by this method depends on the different **electrophoretic mobility** values of the components at a fixed pH and ionic strength (*contrast* **isoelectric focusing**).

If a mixture of proteins A and B is applied at the top of, for instance, a polyacrylamide gel at pH 9 (*see* Figure 62) and an electric field applied, the negatively charged proteins will start to migrate towards the positive pole. If they have similar relative molecular masses and shapes, they will be separated exclusively in respect of their difference in net charge. After a given time, A will migrate in front of B and separation can be achieved. During electrophoresis there is a continuous diffusion of the migrating zones (which was originally fairly narrow); as in chromatographic separations, the longer the migration the greater will be the diffusion.

The different experimental techniques available depend on the nature and the state of dispersion of the material to be separated.

Moving boundary or Tiselius method. This method, which is similar to the moving boundary method for the determination of a **transport number**, is normally used for the characterization and isolation of proteins from biological material. The apparatus (*see* Figure 63) consists of a rectangular-shaped U-tube, each arm of which is composed of two compartments which can slide

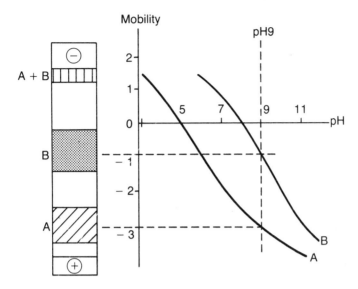

Figure 62. Principles of electrophoretic separation of two proteins A and B

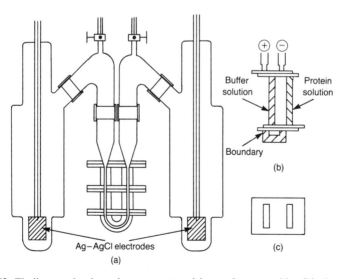

Figure 63. Tiselius moving boundary apparatus: (a) complete assembly; (b) electrophoresis cell; (c) horizontal cross-section of cell

122 Electrophoresis

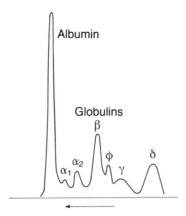

Figure 64. Electrophoretic diagram (ascending) for human blood serum. Schlieren scanning trace. ϕ = fibrinogen

over one another, connected to large electrode compartments containing reversible silver–silver chloride electrodes. A sharp boundary is formed between a solution of the proteins in buffer solution in the lower compartment and buffer solution above. On application of an electrical potential between the electrodes, the different proteins (of different mobility values) move at different rates and the position of the boundaries between the various components is observed optically (by Schlieren scanning, refractive index or interference optics). Measurements are undertaken at constant ionic strength and pH, and at 4°C, to minimize density differences and convection currents. From the recorded traces (see Figure 64) at different times and the known applied potential gradient, the mobility values of the different protein components can be calculated. This is a very accurate method for identifying components in a mixture with the possibility of obtaining samples of the different fractions (by sliding the two compartments of the U-tube apart and removing some of the liquid). It is, however, a time-consuming technique and is unsuitable for rapid screening of biological samples.

Particulate microelectrophoresis. This technique is limited to a study of the electrical properties of particles (colloids, bacteria and other biological cells, oil droplets and gas bubbles) which are microscopically visible. The closed electrophoretic cell (*see* Figure 65(a) and (b)), which may be either cylindrical or rectangular in cross-section, is mounted on the stage of a microscope. This is connected to electrode compartments which are fitted with either silver–silver chloride or platinum electrodes. A suspension of the particles in buffer solution of constant ionic strength and pH is introduced into the cell and the rate of migration of an individual particle, located at the stationary level in the cell, under a known applied field strength is determined by measuring the

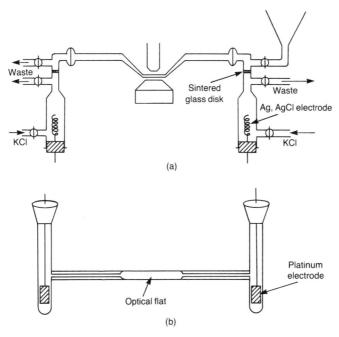

Figure 65. Essential features of electrophoresis assembly: (a) rectangular cell mounted in horizontal position; (b) cylindrical cell for use in lateral position

transit time across a calibrated eyepiece graticule. Such measurements lead to a value for the mobility of each particle in suspension and hence to a distribution of mobilities in the population, or to an average value for a homogeneous population.

The technique is rapid and can be used with suspension media over a wide range of ionic strength and pH. High-power magnification is used and thus the shape, size and orientation of particles and the presence of a mixed population can be observed in an environment which is constant with respect to pH and ionic strength over the short time (less than 10 seconds) required for a measurement.

Zone electrophoresis. Many of the difficulties associated with the Tiselius method can be avoided if the separation is carried out in a stabilizing or supporting medium such as filter paper, cellulose acetate, silica gel or a column packed with a suitable powder. In principle, a potential gradient is applied to the ends of the supporting medium impregnated with a suitable buffer solution; near the centre of which is a drop of the sample under investigation (*see* Figure 66). The differently charged components move with different velocities and form discrete zones which can be located by suitable

124 Electrophoresis

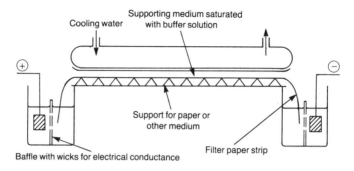

Figure 66. Typical apparatus for paper or zone electrophoresis

means (e.g., staining with specific reagents). Migration depends on: (1) the properties of the ion, its sign, net charge, shape and size, and the state of dispersion (colloidal or molecular); (2) the properties of the buffer solution, pH, ionic strength, temperature and viscosity; (3) the texture of the supporting material; and (4) the current density and field strength. The supporting medium, acting as a series of capillaries, offers a certain resistance to migration, and this with the effects of electroosmosis of the buffer solution through the negatively charged supporting medium means that the electrophoretic mobility (calculated from the apparent distance moved by a component in a measured time and the applied field strength) is less than that obtained from the moving boundary method.

The temperature must not be allowed to rise, otherwise evaporation of solvent and capillary movement will interfere with the electrophoretic separation; commercially available apparatus often includes a cooling system. The time required for a separation depends on the voltage applied, ranging from 18–24 hours for low voltage to 45–60 minutes for high voltages. This method is highly suitable and widely used for the rapid separation and identification of proteins in biological fluids.

Zone electrophoresis carried out in open tubular glass capillaries, which effectively dissipate the heat generated at 30 kV, has a separation efficiency in excess of 400 000 theoretical plates. The strong electroosmotic flow present in this system enables the separation and analysis of both positive and negative ions in a single run.

A variation of zone electrophoresis is the two-dimensional electrochromatography in which the paper is held vertically and the sample components are carried down the sheet by a flow of buffered solvent. As a result of differences in distribution of the components between the fixed and the mobile phases, separation occurs in the vertical direction. Additionally an electrical field is applied horizontally between electrodes mounted along the vertical edges of the paper; thus electromigration occurs along the horizontal

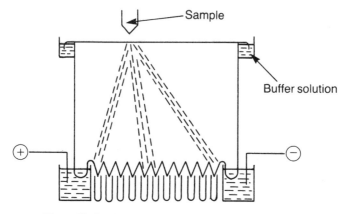

Figure 67. Continuous electrophoretic separation

axis. In consequence, the various components in the sample describe a radial path from the origin (*see* Figure 67). In the preparative mode, samples of the separated species can be collected in tubes.

Further reading
M. Bier (ed.), *Electrophoresis: Theory, Methods and Application*, Academic Press, New York (1959)
A. M. James, Electrophoresis of particles in solution, *Surface Colloid Sci.*, **11**, 121 (1979)
C. F. Simpson and M. Whittaker (ed.), *Electrophoretic Techniques*, Academic Press, New York (1983)

Electrophoretic deposition of polymers and paints. A common method of coating metal parts with a protective film evenly and efficiently. If a pigment such as titanium dioxide is added to the colloidal solution of the coating bath in the ratio of one part pigment to two parts binder a paint may be deposited. The coating is formed at a potential difference of 100–400 V with the part that is to be coated as the anode or cathode, depending on the nature of the binder. Anodic deposition is usually based on the incorporation of the carboxylate anion as part of the polymer or surfactant. The anion is thought to attach itself to an anodized metal surface. In cathodic systems, the carboxylate anion is replaced by an amino or sulphonium cation. Epoxy esters may be coated cathodically with a wide variety of pendant groups (e.g., amines and mercaptans). As in the case of **electroplating** of metals, pretreatment of the electrode is important to the successful coating process. All deposits are baked to produce the final stable coat.

Electrophoretic mobility (u; units: $m^2 s^{-1} V^{-1}$). For a charged particle (protein or colloidal particle, biological cell), the electrophoretic mobility is the linear velocity for unit field strength. Although it is related to the ζ-**potential**, most workers prefer to discuss experimental results in terms of the experimentally determined mobility values under constant environmental conditions (pH, ionic strength, etc.). This avoids making assumptions about the values of the dielectric constant (*see* **Permittivity**) and coefficient of viscosity to be used in the equation (Eq. (132)).

The method for measuring the mobility depends on the state of dispersion of the charged particles (*see* **Electrophoresis**).

Although there is a large difference in size and in the charge-determining species among ions, colloidal particles, biological cells etc. (*see* Table 13), nevertheless the mobility values are all of the same order of magnitude; thus one single mobility value will not serve to characterize a particular species or a particular surface group.

Table 13
Electrophoretic mobility of a range of particles measured at pH 7.0

Species	$10^8 \times$ mobility/$m^2 s^{-1} V^{-1}$
Staphylococcus aureus	
Methicillin-sensitive	−1.00
Methicillin-resistant	−1.48
Streptococcus pyogenes	
Type 2 G	−1.03
Type 2 M	−0.89
Human blood cells	
Erythrocytes	−1.08
Lymphocytes	−1.09
Platelets	−0.85
Hamster kidney	
Erythrocytes	−1.35
Tissue cells	−0.65
Tumour cells	−1.2
Chlorella cells	−1.70
Hydrogen ion	+36.70
Chloride ion	−6.8
Sodium ion	+5.2
Colloidal gold	−3.2
Oil droplets	−3.1

The surface ionogenic groups of many biological cells have been characterized and often identified from studies of the variation of the mobility value with: (1) the nature and concentration of the suspension solution at constant pH; (2) the pH of the suspension medium at constant ionic strength before and after mild chemical or enzymic treatment; and (3) the presence of surfactants in the suspension medium. Changes in the surface properties of bacterial cells accompany changes in their antibiotic resistance pattern and changes in the environmental growth conditions.

Further reading
A. M. James, Molecular aspects of biological surfaces, *Chem. Soc. Rev.*, **8**, 389 (1979)
A. M. James, The electrical properties and topochemistry of bacterial cells, *Adv. Colloid Interface Sci.*, **15**, 171 (1982)

Electroplating. The **electrodeposition of metals** usually in a layer 0.01–0.1 mm thick, for decoration or protection. The greatest industrial consumption of metals used in plating is of tin and nickel, followed by chromium. The article to be plated need not necessarily be metallic; if not, a conducting layer must be formed on it before plating: for example, by applying a graphite or metal powder, or by treating it with silver nitrate and reducing solution, or by direct metallizing in a vacuum.

Drastic cleaning of the article is essential, and may include degreasing in an organic solvent, treatment in alkaline detergent solution, electrolysis for a brief period at high current density, and finally an acid dip. Rinsing is needed between each stage, and in large-scale practice the article may move automatically through the necessary tanks.

The composition of the electrolyte, the current density and other conditions must be adjusted to give a uniform and perfectly adherent deposit of very fine crystalline grain. Adherence is sometimes improved by first depositing a very thin layer of a metal which forms a solid solution with both the underlying metal and the metal to be subsequently deposited; a microcrystalline deposit is usually favoured by the use of a complex salt as electrolyte, complex cyanides being used in most cases. Plating baths may also include buffering agents, small additions of surface active agents found empirically to have a good effect on the structure of the deposit and inert electrolytes. Brightening agents also improve the appearance of the plate: for example, addition of imidazole and thiourea to copper plating baths gives an exceptionally bright plate. It is thought that the action of brighteners is through adsorption at high surface energy points, which suppress dendrite growth and lead to a smoother deposit. A bath should have good 'throwing power': that is it should give a deposit of uniform thickness even when the article has projections (which are closer to the anode) or recesses (where the current density is likely to be reduced). Inert electrolytes have an influence here in determining the proportion of material brought to the surface by conduction; the variation of **overpotential** with current density, and the diffusion rates and chemical stability of the various complex ions present in the surface layer are all factors in determining throwing power. Renewal of the electrolyte concentration at the cathode surface may be left to convective mixing and to the stirring effect of any hydrogen evolved, but additional stirring is usually provided either mechanically or by streams of air pumped into the bath. The anode is usually made of the metal being plated, and if the **current efficiency** is the same for cathode deposition and anode dissolution, the composition of the electrolyte will be maintained. The anode must not be allowed to become passive as, for instance, a nickel anode tends to do at high current density; additions to the bath or to the anode itself may be made to prevent this.

The dissolution of the anode tends to produce an 'anode sludge' ('anode slime'), consisting of insoluble particles of the metal and impurities. These must not be allowed to contaminate the cathode surface, and so the anode is contained in a porous bag or the bath liquid is continuously filtered. The composition of the bath should be kept under constant control, since any evolution of hydrogen at the cathode, or oxygen at the anode, will alter the pH. Any variations in current efficiency may alter the concentration of the cation being discharged, and a low concentration of an additive may gradually disappear through side reactions or by inclusion in the electrodeposit. See Kirk-Othmer (1981).

Further reading
AES, *Symposium on Difficult to Plate Metals*, vol. I, American Electroplaters' Society, Winter Park, Florida (1980)
AES, *Symposium on Difficult to Plate Metals*, vol. II, American Electroplaters' Society, Winter Park, Florida (1982)

Electrorefining. If two electrodes of a metal, such as copper, are immersed in a solution of one of its salts, and a small potential difference is applied between them, metal will dissolve from the anode, the more positive electrode, and deposit in corresponding amount on the cathode. The net chemical reaction is nil, and energy is required only to overcome the resistance of the solution and the (usually very small) **overpotential** at the electrodes.

The principle is very widely used in electrorefining. The impure metal is used as anode, and pure metal is obtained at the cathode. Any impurity metals with a more positive electrode potential than that of the metal being refined will remain undissolved at the anode, and may be recovered as 'anode sludge'. Any metal with an $E^{\ominus}$ more negative than that of the metal being refined will dissolve with it, but the cathode potential will be too low for its discharge and it will accumulate in the bath, and must eventually (or continuously in a circulating system) be removed for chemical treatment.

The conditions for a successful process include: (1) the conductance of the electrolyte should be high, to reduce the *IR* voltage drop; (2) the electrolyte should be selected with two further objectives in mind—it should be such as to give an adherent deposit, and it should give satisfactory dissolution of the anode (preventing any danger of this electrode becoming passive); (3) there must be no danger of a constituent of the anode sludge being taken up into solution by the electrolyte, as it would then be discharged at the cathode; (4) the current density should preferably be high, but must not approach the limiting current density (*see* **Diffusion-limited current**) or lead to the discharge of hydrogen.

The method is very extensively used for obtaining pure copper, for the refining of silver and gold, and as the final stage in the purification of many other metals.

Electrostriction. The volume change or compression caused by an electric field, especially an inhomogeneous electric field which tends to concentrate dipoles in regions of highest field strength. For an ion such as the sodium ion, the local field (approximately 1.3×10^8 V cm^{-1}) is very inhomogeneous, falling off rapidly with the square of the distance from the central ion. Near the ion there is therefore a large tendency to orient solvent dipoles and to concentrate them as near as possible to the ion to minimize their energy. The net result is a tension in the immediate vicinity of the ion which is equivalent in its effect on the local solvent density and structure to the application of a very high pressure. The volume change associated with this tension is known as electrostriction.

Usually non-aqueous, non-associated polar solvents (no hydrogen bonding), such as acetonitrile and propylene carbonate, are more compressible than water, and electrostriction can therefore be larger than in relatively compressible liquids, since compression must bend or break intermolecular hydrogen bonds in the structure with the expenditure of energy. See Conway (1981).

Electrowinning. A process closely related to **electrorefining** in which the anode of the metal being refined is replaced by an insoluble anode: for example, oxygen may be evolved at a lead or lead–silver alloy electrode. Only ions initially present in the solution are plated at the cathode. Zinc is produced largely by electrowinning from acidified zinc sulphate solution. Copper, nickel and cobalt may also be produced by this method. See Bockris, Conway and Yeager (1981), vol. 2.

Ellipsometry. Elliptically polarized light undergoes a change on reflection from a surface which depends on the nature of that surface. Electrode surfaces may be observed *in situ* without exposing them to severe conditions (e.g., vacuum, heat, electron bombardment). By measurement of the change in the amplitude and phase of two orthogonal components of the light, the real and imaginary parts of the reflective index of a clean surface or the thickness and refractive index of a film-covered surface may be determined. Resolution of 2 nm is possible in the measurement of film thickness. Electrochemically generated **oxide films** have been studied using this technique on metals that passivate at anodic potentials. Corrosion processes may also be followed by ellipsometry: for example, the dissolution of passive films of iron shows an initial *cathodic* current coincident with removal of the passive layer. The corrosion process then proceeds on the bare iron surface with an apparent increase in film thickness due to roughening.

Energy density (units: $W\,h\,kg^{-1}$ or $W\,h\,dm^{-3}$). The stored energy in a cell or battery per unit weight or per unit volume. Typical experimental values for common designs of battery are given in Table 14.

Table 14
Values of the energy density of some common cells

Cell	Energy density $/W\,h\,kg^{-1}$	$/W\,h\,dm^{-3}$
Lead–acid	22–33	49–83
Nickel–zinc	37–77	67–134
Nickel–cadmium	24–33	61–90
Silver–zinc	55–220	80–610
Cadmium–air[a]	80–90	14–24
Zinc–air[a]	155–175	
Sodium–sulphur	750 (theoretical)	
Zinc–carbon (**Leclanché**)	55–77	120–152

[a] *See* **Metal–air battery**.

Enzyme electrode. *See* **Biocatalytic membrane electrode.**

Equilibrium constant (K). Many chemical reactions do not go to completion but proceed to a position of dynamic equilibrium where the reaction apparently ceases, often leaving considerable amounts of unchanged reactants. Under defined conditions of temperature and pressure there is a fixed relationship between the active masses of reactants and products. For the general reaction

$$aA + bB + \cdots + iI \rightarrow lL + mM + \cdots + tT$$

the free energy change is given by the van't Hoff isotherm

$$\Delta G = \Delta G^\ominus + RT \ln \frac{[L]^l[M]^m \cdots [T]^t}{[A]^a[B]^b \cdots [I]^i}$$

where the brackets [ ] represent the active masses of reactants and products. When the system is in equilibrium, $\Delta G = 0$ and so

$$\Delta G^\ominus = -RT \ln \frac{[L]^l[M]^m \cdots [T]^t}{[A]^a[B]^b \cdots [I]^i} = -RT \ln K$$

Replacing active masses by activities

$$K_{therm} = \frac{a_L^l a_M^m \cdots a_T^t}{a_A^a a_B^b \cdots a_I^i} = \exp\left(-\frac{\Delta G^\ominus}{RT}\right)$$

Since $\Delta G^\ominus = -nFE^\ominus$, it follows that

$$\ln K_{therm} = \frac{nFE^\ominus}{RT}$$

For ionic equilibria, equilibrium constants can be determined by electrical methods.

1) **Conductiometric methods** for an acid **dissociation constant** and **solubility product**.
2) **Electrometric titrations** for acid dissociation constants and stability constants.
3) **Reversible galvanic cell** for acid dissociation constants, solubility products and the **ionic product of water**.
4) **Concentration cell** for solubility products.

See James (1976); James and Prichard (1977).

Exchange current density. The anodic and cathodic current density when no net current flows at an electrode. This is illustrated in Figure 68, where the cathodic and anodic currents, together with the total current, are plotted for a one-electron reaction. The exchange current density is a function of the concentration of the oxidized and reduced species, temperature and the nature of the electrode material. In the derivation of the **Butler–Volmer equation**, the exchange current density (i_0) is given by

$$i_o = a_{ox} F\left(\kappa_f \frac{kT}{h}\right) \exp\left(-\frac{\Delta G_f^\ddagger}{RT}\right) \exp\left(-\frac{\beta F}{RT}\phi_{m,e}\right)$$

$$= a_{red} F\left(\kappa_b \frac{kT}{h}\right) \exp\left(-\frac{\Delta G_b^\ddagger}{RT}\right) \exp\left[\frac{(1-\beta)F}{RT}\phi_{m,e}\right] \quad (67)$$

132 Exchange current density

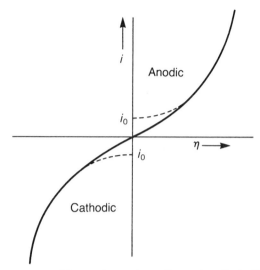

Figure 68. Current–overpotential curve for an electrodic reaction. Dashed lines are the anodic and cathodic currents. Full line is the net current

From Eq. (67), if $\kappa_f = \kappa_b = 1$

$$\phi_{m,e} = \frac{RT}{F}\left[-(\Delta G_f^{\ddagger} - \Delta G_b^{\ddagger}) + \ln \frac{a_{ox}}{a_{red}}\right] \quad (68)$$

$$= \phi_m^{\ominus} + \frac{RT}{F} \ln \frac{a_{ox}}{a_{red}} \quad (69)$$

Eq. (69) is the **Nernst equation**. A rate constant (k^0) is defined as

$$k^0 = \frac{kT}{h} \exp\left(-\frac{\Delta G_f^{\ddagger}}{RT}\right) \exp\left(-\frac{\beta F \phi_m^{\ominus}}{RT}\right)$$

$$= \frac{kT}{h} \exp\left(-\frac{\Delta G_b^{\ddagger}}{RT}\right) \exp\left[-\frac{(1-\beta)F\phi_m^{\ominus}}{RT}\right] \quad (70)$$

Therefore

$$i_0 = Fk^0 a_{ox}^{1-\beta} a_{red}^{\beta} \quad (71)$$

In general for a multistep reaction

$$i_0 = nFk^0 a_{ox}^{1-\alpha} a_{red}^{\alpha} \quad (72)$$

Exchange current densities cannot be measured directly, and are usually obtained from plotting overpotential–current data as the **Tafel equation**. *See also* **Electrocatalysis; Transfer coefficient.**

F

Falkenhagen effect. *See* **Conductance at high frequencies.**

Farad (F; $F = C V^{-1} = A s V^{-1} = A^2 s^4 kg^{-1} m^{-2}$). SI derived unit of electrical capacitance. A capacitance of 1 farad (1 F) requires 1 coulomb of electricity to raise its potential by 1 volt.

Faraday constant (F). Charge carried by 1 mole of electrons, that is

$$F = \text{electronic charge} \times N_A$$
$$= 1.602 \times 10^{-19} \text{ C} \times 6.022 \times 10^{23} \text{ mol}^{-1}$$
$$= 96\,472 \text{ C mol}^{-1}$$

A value of $96\,490 \pm 3$ C mol^{-1} has been obtained experimentally using a silver coulometer.

Faraday's laws. Laws of electrolytic conduction.

1) In any electrolytic process, the amount of chemical reaction is proportional to the quantity of electricity passed through the electrolytic conductor.
2) The masses of different substances deposited or dissolved by the same quantity of electricity are in the ratios of their chemical equivalents.

In ideal cases, these laws are exact, although this fact may be obscured in certain cells by the occurrence of side reactions.

Taken together these laws mean that a certain quantity of electricity will theoretically discharge one gram equivalent of any substance. This quantity of electricity is the Faraday constant F and is given by

$$F = \frac{QE}{w} = \frac{ItA}{wz}$$

where Q is the quantity of electricity which produced w g of a substance of equivalent weight E, that is A/z where A is the relative atomic mass and z the valence of the element deposited. The electrochemical equivalent of an element is the weight deposited or dissolved by 1 C in an electrolytic process (i.e. when $Q = 1$ C), then the electrochemical equivalent $w = E/F = A/zF$.

Fick's laws of diffusion

Fick's laws of diffusion. The rate at which a substance diffuses across a unit cross-sectional area depends not only on the molecular shape and size but also on the concentration gradient ($\partial c/\partial x$) of that substance. In the absence of other influencing factors, matter moves spontaneously from a region of higher to one of lower concentration. Fick's first law states

$$\frac{\partial m}{\partial t} = -DA\left(\frac{\partial c}{\partial x}\right)_t \qquad (73)$$

where $\partial m/\partial t$ is the rate at which the substance crosses an area A and D is the diffusion coefficient of the solute. Alternatively Eq. (73) may be written in the form

$$J = \frac{(\partial m/\partial t)}{A} = -D\left(\frac{\partial c}{\partial x}\right)_t \qquad (74)$$

where J is the flux, that is the increment ∂m crossing unit area in unit time. If J is expressed in mol m^{-2} s^{-1} and the concentration gradient in mol m^{-4}, D has the units of m^2 s^{-1} and is independent of the units of mass chosen. For KCl at 25°C, $D = 1.99 \times 10^{-10}$ m^2 s^{-1} and for sucrose $D = 5.23 \times 10^{-10}$ m^2 s^{-1}.

Fick's second law states

$$\left(\frac{\partial c}{\partial t}\right)_x = D\left(\frac{\partial^2 c}{\partial x^2}\right)_t \qquad (75)$$

This gives an expression for the rate of change of concentration at a point x from the origin in terms of the diffusion coefficient and the change of concentration gradient in the direction of diffusion at time t.

Fixed bed electrode. A three-dimensional electrode in the form of a porous, high surface area powder. The structure of a fixed bed electrode differs from that of a **porous electrode** by providing extra channels for the flow of electrolyte between the particles of the bed, allowing higher rates of flow and reducing the number of dead-end pores. The construction of a cell containing a fixed bed electrode is shown in Figure 69. The high surface area of a fixed bed electrode, and its efficient operation in a flow system, leads to its use in the recovery of metals from very dilute solutions. A bed of graphite powder has been shown to be able to remove copper ions from solutions containing 10^{-4} mol dm^{-3} Cu^{2+}. The plated graphite is anodically redissolved in a much smaller volume of electrolyte. This leads to a considerable concentration of the metal ion. The method is akin to a large-scale anodic stripping experiment (*see* **Stripping voltammetry**). See also **Fluidized bed electrode**.

Flade potential. *See* **Passivity**.

Flemion membrane. *See* **Ion-selective membrane**.

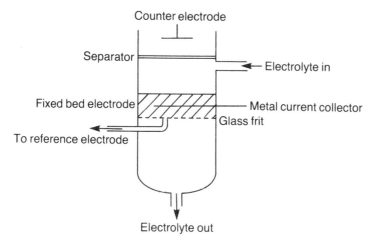

Figure 69. Fixed bed electrode in a cell with flowing electrolyte

Fluidized bed electrode. A bed of discrete, electronically conducting particles supported against gravity by the flow of electrolyte. The form of a fluidized bed electrode is similar to that of a **fixed bed electrode** (*see* Figure 69) except that the flow of electrolyte is reversed. A high ratio of electrode area to cell volume is achieved with good heat and mass transfer characteristics. Ohmic losses down the bed, both in electrolyte and electrode, give an unwanted variation in **potential distribution**. Fluidized bed electrodes may be used for recovery of metals from dilute solutions and for **organic synthesis**.

Formal electrode potential (E'). *See* **Redox electrode systems.**

Formation constant. Many metal ions react with electron pair donors to form coordination compounds or complex ions. The donor species or ligand (L) must have at least one pair of unshared electrons available for bond formation. Common ligands are water, ammonia, amino acids and halide ions. For example

$$M + nL \rightleftharpoons ML_n$$

where the charges on M and L determine the charge on the product. The formation constant for this equilibrium is given by

$$K_f = \frac{a_{ML_n}}{a_M a_L^n}$$

The formation constant for such a complex can be obtained from measurements of the e.m.f. of a **concentration cell** without transport. Thus for the equilibrium

136   Frequency dependence of conductance

$$Ag^+ + 2NH_3 \rightleftharpoons Ag(NH_3)_2^+$$

$$K_f = \frac{a_{Ag(NH3)2+}}{a_{Ag+} a_{NH3}^2} \approx \frac{c_{Ag(NH3)2+}}{c_{Ag+} c_{NH3}^2}$$

(assuming the $\gamma_{Ag(NH3)2+} = \gamma_{Ag+}$ and $\gamma_{NH3} = 1$) a suitable cell would be

Ag | 0.025 mol dm^{-3} AgNO$_3$ in x mol dm^{-3} NH$_3$ | saturated KNO$_3$ | 0.01 mol dm^{-3} AgNO$_3$ | Ag
$(a_{Ag+,1})$ $\qquad\qquad\qquad\qquad\qquad\qquad\qquad\qquad\qquad\qquad\quad$ $(a_{Ag+,2})$

the e.m.f. of which is given by

$$E = \frac{RT}{F} \ln \frac{c_{Ag+,2}}{c_{Ag+,1}}$$

From a knowledge of the concentration of silver ions ($c_{Ag+,2}$) in the right-hand half cell, the concentration of silver ions ($c_{Ag+,1}$) can be calculated and hence, by difference, the concentration of the complex.

Formation constants can also be obtained by potentiometric titration (*see* **Electrometric titrations**) to determine the equilibrium concentration of free ligand or free cations for known initial concentrations of cation and ligand. See James and Prichard (1977).

Frequency dependence of conductance. *See* **Conductance at high frequencies.**

Fuel cell. A cell in which the reactants of an exoenergetic reaction are continuously supplied, and the products continuously removed, so that electricity may be supplied indefinitely. The main attraction of a fuel cell, compared with other power sources, is the high efficiency that can be theoretically attained in utilizing the free energy change of a reaction as available energy (*see* **Electrochemical energy conversion**). Some possible fuels are given in Table 15. To realize this high efficiency in practice requires that

Table 15
Theoretical e.m.f. and efficiency values of some possible fuel cell reactions

Reaction	$E^{\ominus}$/V	$\varepsilon = \Delta G^{\ominus}/\Delta H^{\ominus}$
$H_2 + \frac{1}{2}O_2 \rightarrow H_2O$	1.229	0.83
$CH_4 + 2O_2 \rightarrow CO_2 + 2H_2O$	1.060	0.92
$CH_3OH + \frac{3}{2}O_2 \rightarrow CO_2 + 2H_2O$	1.206	0.97
$N_2H_4 + O_2 \rightarrow N_2 + 2H_2O$	1.559	0.99
$CO + \frac{1}{2}O_2 \rightarrow CO_2$	1.332	0.91

the internal resistance of the cell and the **overpotential** at each of the electrodes be low. So far the most highly developed cell is the hydrogen–oxygen gas cell. The theoretical voltage that can be obtained from the combustion of hydrogen in oxygen is 1.23 V at 298 K, corresponding to the reaction

$$\text{cathode} \quad \tfrac{1}{2}O_2 + 2H^+ + 2e \rightarrow H_2O$$
$$\text{anode} \quad H_2 \rightarrow 2H^+ + 2e$$
$$\text{overall} \quad H_2 + \tfrac{1}{2}O_2 \rightarrow H_2O$$

With efficient catalysts, this reaction can give rise at ordinary temperatures to current densities of 150–200 mA cm^{-2} at about 0.75 V. Medium temperature cells operating at 370–470 K, such as the Bacon cell, can give up to 1000 mA cm^{-2} at the same voltage and with less expensive catalysts. The main problem in design is to bring about the electrode reactions sufficiently rapidly and at low overpotentials. This is partly solved by using electrode surfaces that catalyse the reaction efficiently (*see* **Electrocatalysis**). The oxygen reaction, particularly, has a low **exchange current density** and is normally very slow (*see* **Oxygen electrode reactions**). The other side to the problem is to bring the gaseous reactants to the electrode/electrolyte interface sufficiently rapidly to maintain reasonably high currents. This is done by using electrodes consisting of thin sheets of porous material (metal, carbon or metal-coated plastic) and arranging the gas pressure and the pore size so that the gas/liquid interface is within the pores of the material. The liquid meniscus within each pore thus provides an area of electrode wetted by a very thin film of liquid through which the reactant can reach the active surface very rapidly (*see* **Porous electrode**).

The electrolyte is normally a concentrated solution of a strong acid or alkali; this will have a high conductance and will help to minimize the concentration overpotential. Potassium hydroxide is the first choice as it is less corrosive than concentrated acids, but its usefulness is limited because of its reactivity towards carbon dioxide. If used in a hydrogen–air cell, the air must be purified, and for any cell designed to use a hydrocarbon fuel potassium hydroxide would be unsuitable as carbon dioxide would be one of the products of the cell reaction. The alternatives are concentrated sulphuric acid, which is only suitable at moderately low temperatures, and phosphoric acid.

The thickness of the electrolyte layer may be as little as 1 mm. In the fuel cells used in the Gemini spacecraft, the electrolyte was replaced by an ion-selective membrane. This provided a conducting bridge between the electrodes, and contributed to the mechanical strength; it also supplied a source of drinking water, since the product of the cell reaction (about half a litre of water per kilowatt hour) could be collected in a pure state.

As already indicated, a flat layer construction is normally used in fuel cell batteries, and an individual cell, usually not more than 1 cm thick, will be

138   Fuel cell

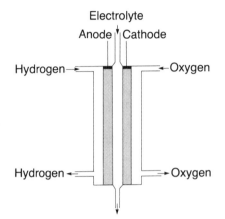

Figure 70. Hydrogen–oxygen fuel cell (schematic)

somewhat as shown in Figure 70. Low-temperature cells need the most effective catalyst, which is platinum, although silver or some transition metal oxides can be used at the oxygen electrode. To prevent the cost being prohibitive, efforts are made to restrict the platinum coating to the actual region of contact between gas and liquid. In medium-temperature cells, such as the Bacon cell, which operate at about 470 K, nickel is used as catalyst at the hydrogen electrode and nickel oxide (on a base of porous nickel) at the oxygen electrode.

The ancillary equipment for such a cell includes pumps, to circulate the gaseous reactants, a condensing unit to remove the water produced from the gas streams, and a heating unit to provide the starting temperature (enough heat is generated in the working cell to maintain this). Another practical factor is the source of hydrogen. Electrolytic hydrogen will not come under consideration, except possibly in electricity storage systems, or where cost is unimportant. It could be produced *in situ* by the action of steam on a cheap liquid fuel, by reactions such as

$$C_3H_8 + 6H_2O \rightarrow 10H_2 + 3CO_2$$
$$CH_3OH + H_2O \rightarrow 3H_2 + CO_2$$

or from liquid ammonia by a catalysed thermal decomposition

$$2NH_3 \rightarrow N_2 + 3H_2$$

The latter has the advantage that nitrogen is innocuous, whereas carbon dioxide might have to be removed. There would be obvious advantages in using a cheap, freely available organic fuel directly in the cell, and avoiding the subsidiary reactions just mentioned. Methanol, dissolved in the electrolyte, or propane will react anodically

$$CH_3OH + H_2O \rightarrow 6H^+ + CO_2 + 6e$$
$$C_3H_8 + 6H_2O \rightarrow 20H^+ + 3CO_2 + 20e$$

These reactions are slow and their useful application requires expensive porous metal catalysts and temperatures above ambient. Alkaline electrolytes may be used with some organic fuels, but, as shown above, there is always a problem with dissolved carbon dioxide. Organic fuel cells must also have separate anode and cathode compartments, as most organic fuels polarize the air cathode leading to a loss of fuel and power. At present, the disadvantages in the direct use of such fuels appear to be great, and outweigh the extra cost of reforming them to produce hydrogen for a Bacon cell.

A cell running at a high temperature will, in terms of the electrochemical reactions, operate at greatly increased efficiency. This is offset by the considerable problems of materials technology associated with the construction and operation of the cell. A molten carbonate cell operating at 500°C (the eutectic of sodium carbonate and lithium carbonate) has the reactions

cathode $\quad \frac{1}{2}O_2 + CO_2 + 2e \rightarrow CO_3^{2-}$
anode $\quad\quad H_2 + CO_3^{2-} \rightarrow CO_2 + H_2O + 2e$

overall $\quad\quad H_2 + \frac{1}{2}O_2 \rightarrow H_2O$

although a hydrocarbon fuel, whose oxidation products include carbon dioxide, may be preferable. A diagram of a molten carbonate cell is shown in Figure 71.

A cell which operates at even higher temperatures (>1000°C) uses the conducting properties of doped zirconium oxide. For example, at 1000°C 85 percent zirconium oxide/15 percent copper (II) oxide has a conductivity of

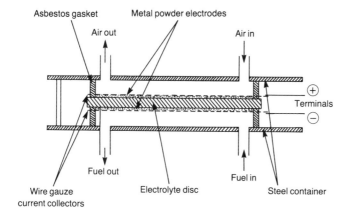

Figure 71. A molten carbonate fuel cell

$2 \times 10^{-2} \, \Omega^{-1} \, cm^{-1}$, owing to the movement of oxide ions through the lattice (*see* **Solid electrolytes**). Silver is a good electrode material, and is deposited on a thin disc of electrolyte. In operation, the efficiency of the electrode reaction is high, most of the voltage drop arises from the resistance of the electrolyte.

Further reading
A. McDougal, *Fuel Cells*, Macmillan, London (1976)

Fuoss conductance equation. *See* **Conductance equations.**

Fused salt electrochemistry. Fused salts as solvents and electrolytes are employed when access to potentials outside the limits of stability of water (somewhat more negative than 0 V versus standard hydrogen electrode and more positive than 1.23 V versus standard hydrogen electrode) are required to effect electrochemical reactions of ions. A major difficulty lies in the high temperatures required to form the melts and the materials problems associated with this. As an example of a fused salt system the lithium chloride–potassium chloride eutectic (which contains 59 percent lithium chloride) which melts at 450°C is chosen to illustrate some of the features of fused salt electrochemistry.

Potential range. The limiting reactions of the melt are chlorine evolution and lithium deposition which occur at $+1.033$ V (versus Ag (I)| Ag) and -2.593 V, respectively, giving a working range of 3.6 V.

Reference electrodes. The two most popular reference electrodes in this system are the Pt (II)|Pt electrode and the Ag (I)|Ag electrode. The chlorine electrode ($Cl_2|Cl^-$) is also suitable, but suffers from the practical problems associated with bubbling chlorine through the melt.

Electrochemical systems
1) Oxygen. An oxygen electrode of the type $O_2|M|O^{2-}$ has not been successfully developed in this melt. However, electrodes $M|M_xO_y|O^{2-}$ have been studied where M is platinum, palladium or bismuth. Hydroxide ions are probably not stable in the melt and dissociate to give O^{2-} and water (which boils off).
2) Hydrogen. This exists as a component of a molecular species (e.g. hydrogen chloride) or as the hydride ion H^-, which may be produced by the addition of lithium hydride. The reactions of hydrogen chloride have been suggested as the basis of an industrial process for the conversion of hydrogen chloride to hydrogen and chlorine

$$HCl + e \rightarrow \tfrac{1}{2}H_2 + Cl^- \qquad (a)$$
$$Cl^- \rightarrow \tfrac{1}{2}Cl_2 + e \qquad (b)$$

Table 16
Electrode potentials of halogen electrodes in the lithium chloride–potassium chloride eutectic melt at 450°C

Electrode	$E^\ominus$ (versus Ag(I)|Ag)/V
$F^-\|F_2(g),C$	Not stable in this melt
$Cl^-\|Cl_2(g),C$	1.033
$Br^-\|Br_2(g),C$	0.857
$I^-\|I_2(g),C$	0.473

The reverse process of reactions (a) and (b) would make a useful **fuel cell** reaction.

3) Halogens. Apart from fluorine, whose electrochemistry lies outside the range of potentials in which the melt is stable, electrode potentials of the halogens have been measured (*see* Table 16). In the case of chlorine, the trichloride ion (Cl_3^-) has been identified, and it is thought that this ion may play a role in determining the electrode potentials associated with reactions of chlorine.

4) Metals. The electrode potentials for many metal/metal ion half cells have been measured. Of interest are metals which may be plated from the melt. These include ruthenium, iridium, rhodium, cobalt, copper, silver, gold, vanadium and lanthanum. Lithium is deposited from the melt at its cathodic limit. *See also* **Fused salts**. See Bard (1976), vol. X.

Further reading
D. Inman and D. G. Lovering, Electrochemistry in molten salts, in *Comprehensive Treatise Electrochemistry*, vol. 7, J. O'M. Bockris *et al.*, Plenum Press, New York, chapter 9 (1983)

Fused salts. Conductance measurements, although more difficult to compare than in water because of wide differences in temperature and viscosity, have provided data of use in the interpretation of the properties of fused salts.

Since fused salts have a much higher **conductivity** than that of the corresponding aqueous system, capillary cells with large cell constants are used for study. Very pure salts must be used; traces of water and oxygen must be excluded to avoid hydrolysis and oxidation at the high temperatures of measurement. To minimize the effects of polarization, grey platinum electrodes are used and measurements are made over a range of frequency values.

The molar conductivity of a fused electrolyte is expressed in terms of the molar volume V_m or density of the melt

142 Fused salts

$$\Lambda = \frac{\kappa}{c} = \kappa V_m = \frac{\kappa M_r}{\rho} \tag{76}$$

where

$$\kappa = \frac{J}{R} = JG$$

(*see* **Conductivity**, Eq. (34)).

The conductivity of molten salts increases markedly with increase in temperature which, since conductance is an activated process of ionic migration under an applied field, can be expressed by an equation of the type

$$\Lambda = A \exp\left(-\frac{E}{RT}\right) \tag{77}$$

where A is a constant and E the electrolyte activation energy of migration (in many systems the separate energies of activation E_+ and E_- are approximately equal). The graph of ln Λ against T^{-1} is linear and of slope $-E/R$. A study of the activation energies of migration has been of assistance in the development of the theory of liquid structure.

Deviations occur at temperatures just above the melting point, where the conductance varies very rapidly with temperature due to the partial collapse of the crystal lattice. The lattice is not completely dispersed until the temperature is well above the melting point (about 10 percent above the melting point, expressed in kelvins). Above this temperature, the increase of conductance with temperature becomes uniform and Eq. (77) is followed. This type of interpretation is supported by X-ray diffraction patterns which change from a well-defined pattern for the crystalline ionic solid, through a blurred pattern near the melting point, to a complete loss of symmetry at higher temperatures.

To overcome differences in viscosity and melting point for the different fused salts it is usual to compare molar conductivities at a temperature 10 percent above the melting point; regarding these as 'corresponding temperatures' (*see* Table 17). The molar conductivities of the alkali halides at temperatures near the melting point are of the same order of magnitude as for their aqueous solutions. There is no doubt that their melts consist essentially of free ions, constituting a new type of solvent in which interionic forces are very high. The results for the alkali metal chlorides differ from those in aqueous solution in that the molar conductivity is highest for the lithium salt, decreasing with increasing crystallographic radius down the group. This is in agreement with the supposition that, in the absence of solvent, it is the bare ion which moves in the applied field. The alkaline earth chlorides are also good conductors, but the order is reversed (i.e. $\Lambda_{BaCl_2} > \Lambda_{MgCl_2} \gg \Lambda_{BeCl_2}$); this is probably due to ion pairing. The very low conductivities of the beryllium and zinc halide melts suggest that they are predominantly polymeric in the molten state, similarly crystalline $AlCl_3$ fuses to a covalent Al_2Cl_6.

Table 17
Molar conductivities of fused metal halides

Salt	Melting point /K	Corresponding temperature /K	Molar conductivity at corresponding temperature $10^4 \Lambda/\Omega^{-1} m^2 mol^{-1}$	Activation energy E/kJ mol^{-1}
LiCl	886	975	183	7.19
NaCl	1074	1181	150	11.29
KCl	1049	1154	120	13.63
RbCl	988	1087	94	15.3
CsCl	919	1011	86	15.68
LiBr	820	902	177	
NaBr	1028	1131	148	10.78
KBr	1003	1103	109	14.30
NaI	924	1016	150	8.36
KI	996	1096	104	15.47
AgCl	728	801	118	5.02
AgBr	707	778	99	4.14
AgI	830	913	105	3.76

Molten nitrates have different characteristics to molten halides; conductance and other studies suggest that ion association is prevalent involving a charge transfer mechanism

$$Na^+NO_3^- + Na^+ \rightleftharpoons Na^+NO_3^-Na^+ \rightleftharpoons Na^+ + NO_3^-Na^+$$

X-ray evidence supports the existence of linear ion triplets. Although the molar conductivities of molten alkali metal hydrogen sulphates are less than those of the corresponding molten alkali metal halides, the activation energies of migration are appreciably higher. This has been attributed to the presence of polymeric hydrogen-bonded structures which restrict the mobility of the anion and so reduce the overall conductance

The high viscosity of the bisulphate melts supports such a suggestion.

It is impossible to determine transport numbers in melts in the same way as in a normal solvent, because no concentration gradients result; the frame of reference provided by a stationary solvent is absent. Mobility values have been derived from self-diffusion measurements using radiotracers. From these results for motion under a known concentration gradient, the mobility under a unit potential gradient can be calculated.

Mixtures of molten salts have the advantage of lower melting points and comparative results can be obtained by using one salt as the main constituent—the solvent—to which smaller amounts of the second substance are added. For mixtures of the salts of the alkali metals, the conductances are approximately additive. When an alkali metal chloride is mixed with a poorly conducting chloride, such as that of zinc, cadmium or lead, there is a minimum in the plot of molar conductivity against the concentration of the melt; this concentration corresponds to the composition of a complex anion. The minimum value is due to the lower mobility of the large anion compared to that of the smaller unassociated ions: for example

$$Cd^{2+} + 2Cl^- + 2(M^+ + Cl^-) \rightleftharpoons 2M^+ + CdCl_4^{2-}$$

where M is potassium, caesium or rubidium. Lithium and sodium chlorides do not form corresponding complex anions with these poorly conducting chlorides, implying that the large anions are stabilized by the large alkali cations but not by the smaller polarizing lithium and sodium ions.

The conductance of metals dissolved in metal halides is high and largely electronic in nature, involving charge transfer between ions and metal atoms: for example, sodium dissolved in fused sodium chloride

$$Na^+ + Na \rightleftharpoons Na^+Na \rightleftharpoons Na + Na^+$$

Faraday's laws will not apply to such a melt.

Further reading

Y. K. Delimarskii and B. F. Markov, *Electrochemistry of Fused Salts*, Sigma (1962)

D. Inman, A. D. Graves and A. A. Nobile, Electrochemistry of molten salts, in *Electrochemistry*, vol. 3, Specialist Reports, Chemical Society, London, p. 61 (1972)

D. Inman and D. G. Lovering. Electrochemistry of molten salts, in *Comprehensive Treatise on Electrochemistry*, vol. 7, J. O'M. Bockris, B. E. Conway and E. Yeager (ed.), Plenum Press, New York, chapter 9 (1983)

G. N. Papatheodoru, Structure and thermodynamics of molten salts, in *Comprehensive Treatise on Electrochemistry*, vol. 5, J. O'M. Bockris, B. E. Conway and E. Yeager (ed.), Plenum Press, New York, p. 399 (1982)

G

Galvanic cell. *See* **Reversible galvanic cell.**

Galvani potential. *See* **Interfacial potential.**

Galvanostat. A device that maintains a constant current through an electrochemical cell. In its simplest form a high, variable resistor is placed in series with the cell. If the variable resistance is of the order of a hundred times that of the cell, small changes in the cell resistance will not affect the total current appreciably. For currents greater than a few milliamps, an electronic circuit is used incorporating transistors and Zener diodes. Alternatively a **potentiostat** may be used in galvanostatic mode.

Gas electrode. An **electrode** in which one part of the electrode couple is a gas, such as hydrogen or chlorine. A chemically inert conductor, usually platinized platinum, is used to adsorb the gas, to transport the electrons and to catalyse the electrode reaction.

The **electrode potential** depends on the pressure of the gas and the activity of the ions in solution. Thus for the chlorine electrode

$$E_{Cl_2,Cl^-} = E^{\ominus}_{Cl_2,Cl^-} + \frac{RT}{F} \ln \frac{p_{Cl_2}^{1/2}}{a_{Cl^-}}$$

This in combination with a **hydrogen electrode** gives the cell

$$\ominus \quad Pt,H_2 \,|\, H^+ \, Cl^- \,|\, Cl_2, Pt \quad \oplus$$

for which the reaction, for the passage of 1 faraday of electricity, is

$$\tfrac{1}{2}H_2 + \tfrac{1}{2}Cl_2 \rightleftharpoons H^+ + Cl^-$$

in which hydrogen chloride is formed at the molality existing in the solution

$$E_{cell} = E_{Cl_2,Cl^-} - E_{H^+,H_2}$$
$$= E^{\ominus}_{Cl_2,Cl^-} - \frac{2RT}{F} \ln a_{\pm} + \frac{RT}{2F} \ln \frac{p_{Cl_2}}{p_{H_2}}$$

See also **Oxygen probe.**

146 Gas-sensing membrane probe

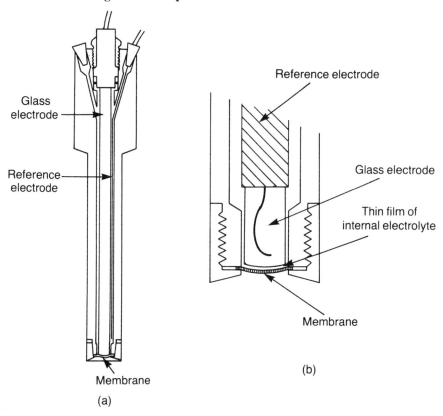

Figure 72. Construction of typical gas-sensing probe: (a) overall layout; (b) cross-section of sensing tip

Gas-sensing membrane probe. Sensor, which is neither ion-selective nor is it an electrode. It consists of an **ion-selective electrode** combined with a suitable reference electrode to form a complete cell. The e.m.f. of the cell is a function of the activity of the gas to be determined.

A typical sensor (*see* Figure 72) is based on a glass electrode with a convex, pH-sensitive tip, held against a gas-permeable membrane (<100 μm thick) to sandwich a thin layer of internal electrolyte between the glass and the membrane. When the tip is immersed in a sample, the gas diffuses through the membrane until the partial pressure of the gas is the same on both sides of the membrane. This equilibrium partial pressure determines the pH of the medium which is measured by the glass electrode and reference silver–silver chloride electrode dipping in the internal electrolyte solution. Since the membrane is hydrophobic, ions cannot enter the probe and therefore will not have a direct effect on the measurement.

The dependence of the e.m.f. generated is related to the concentration of the gas according to the **Nernst equation**

$$E = E' \pm \frac{RT}{F} \ln c_X$$

where the negative sign applies to basic gases (e.g., ammonia) and the positive sign to acid gases (e.g., sulphur dioxide).

The cell must be calibrated at a controlled temperature with solutions of known concentration of the gas and the limits of Nernstian response established.

Gas-sensing probes show outstanding selectivity. The only interfering substances are volatile species which can diffuse into the thin film of electrolyte and which have acidic or basic properties comparable to those of the gas under study.

The response time increases as the lower limit of detection is approached. The pH of the test solution must be adjusted so that all the material to be determined is present as dissolved gas; for ammonia pH > 12 and for sulphur dioxide pH < 0.7.

Gas-sensing probes can conveniently be used in continuous flow analytical systems. *See also* **Ammonia-sensing probe; Sulphur dioxide probe.**

Further reading
M. Riley, Gas-sensing probes, in *Ion-Selective Electrode Methodology*, vol. II, A. K. Covington (ed.), CRC Press, Cleveland, p. 1 (1979)

Gel immunoelectrophoresis. A two-step procedure in which proteins and other antigenic substances are characterized by both their electrophoretic and immunological properties. Proteins, separated by zone electrophoresis (*see* **Electrophoresis**) diffuse into the gel forming precipitation lines with the corresponding antibodies. At the end of the electrophoretic run, specific antiserum is applied to troughs (previously cut in the gel) and the plates incubated in a moist cabinet for immunodiffusion and subsequent precipitation to occur. The plates are finally stained for inspection.

Further reading
C. F. Simpson and M. Whittaker, *Electrophoretic Techniques*, Academic Press, New York, chapter 6 (1983)

Glass electrode. The e.m.f. of the cell

| Ag,AgCl | HCl (0.1 mol dm^{-3}) | glass | test solution | KCl (saturated) | Hg$_2$Cl$_2$ |

←————— glass electrode —————→

Glass electrode

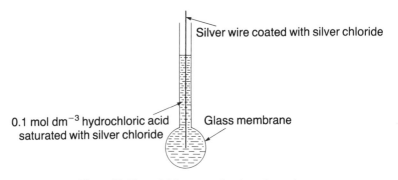

Figure 73. Essential features of a glass electrode

varies with the hydrogen ion concentration of the test solution. The glass electrode may therefore be used for the determination of pH; the **electrode potential** may be represented by the equation

$$E_G = E'_G + \frac{RT}{F} \ln a_{H3O+} \qquad (78)$$

The glass electrode consists of a bulb of special glass blown on the end of a glass tube, containing hydrochloric acid and a reference silver–silver chloride electrode (*see* Figure 73). The electrode should never be allowed to dry or to be used in solutions of pH > 11.

The surfaces of a glass electrode must be hydrated to function as a hydrogen-indicating electrode; hydration is accompanied by an exchange reaction between univalent cations on the glass (Gl^-) and the protons in solution

$$\begin{array}{cccc} H^+ & + \; Na^+Gl^- & \rightleftharpoons \; Na^+ & + \; H^+Gl^- \\ \text{solution} & \text{solid} & \text{solution} & \text{solid} \end{array}$$

The equilibrium constant of this reaction is so large that the surface of a hydrated glass membrane is almost entirely silicic acid (H^+Gl^-). All the singly charged sites of the outer surfaces of the silicic acid gel (about 0.1 μm thick) on both the inside and outside of the glass membrane are occupied by hydrogen ions. Progressing towards the solid glass there is a continuous decrease in the number of protons and an increase in the number of sodium ions (*see* Figure 74).

The potential across a glass membrane consists of a boundary potential and a diffusion potential; only the boundary potential is affected by the pH of the external solution. The potentials E_1 and E_2 are determined by the hydrogen activities of the internal (external) solutions and of the appropriate gel surface.

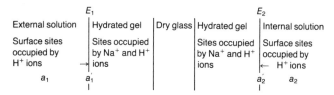

Figure 74. Schematic representation of a glass electrode; a_1 and a_2 are the activities of the hydrogen ion in the external and internal solutions, respectively, and a_1' and a_2' the corresponding activities at the gel surfaces

$$E_1 = \text{constant}_1 + \frac{RT}{F} \ln \frac{a_1}{a_1'} \tag{79}$$

and

$$E_2 = \text{constant}_2 + \frac{RT}{F} \ln \frac{a_2}{a_2'} \tag{80}$$

whence the electrode potential of the glass electrode is given by

$$E_G = E_1 - E_2 = \frac{RT}{F} \ln \frac{a_1 a_2'}{a_1' a_2} \tag{81}$$

If the surfaces of the gels on both sides have been properly prepared, they will have the same number of sites from which protons can leave, then the constants in Eq. (79) and Eq. (80) will be identical and the activities on the gel surfaces (a_1' and a_2') will be identical provided that all the sodium ions have been replaced by protons, thus Eq. (81) becomes

$$E_G = \frac{RT}{F} \ln \frac{a_1}{a_2}$$

but since the activity of the protons on the inside is constant, thus

$$E_G = E_G' + \frac{RT}{F} \ln a_1 \tag{78}$$

where E_G' is a constant.

In alkaline solutions (pH > 10), some glass membranes respond to changes in the concentration of alkali metal ions in solution in addition to changes in the pH; this gives rise to a negative pH error. The error is due to an exchange equilibrium between protons on the gel surface and cations in solution

$$\underset{\text{gel}}{H^+Gl^-} + \underset{\text{solution}}{M^+} \rightleftharpoons \underset{\text{gel}}{M^+Gl^-} + \underset{\text{solution}}{H^+}$$

for which

150 Glucose electrode

$$K_{ex} = \frac{a_1 a'_{M+,1}}{a'_1 a_{M+,1}}$$

where $a_{M+,1}$ and $a'_{M+,1}$ are the activities of the metal cation in solution and on the surface of the gel, respectively. The electrode potential, under these conditions, becomes

$$E_G = E'_G + \frac{RT}{F} \ln (a_1 + K_{ex} a_{M+,1}) \tag{82}$$

Eq. (82) does not adequately describe the electrode potential developed in the presence of other cations; other empirical equations have been advanced.

In use, the glass electrode assembly (i.e. with the reference calomel electrode) must be calibrated using solutions of known pH values (*see* Appendix, Table 7) to eliminate asymmetry potentials in the glass. The accuracy of measurement is ±0.01 pH, or better if extreme care is taken, over the range pH 2–10. It reaches equilibrium immediately in any solution, and its response is unaffected by the presence of any gas, oxidizing and reducing agents, and poisons in the generally accepted sense. It has no appreciable salt or protein error. It is available commercially in many forms for pH measurements in solutions, emulsions or pastes, and for measurements on a micro scale. Combined electrodes in which the glass and calomel electrodes are in a single unit are convenient in that they reduce the size of the measuring unit considerably.

The electrode suffers from asymmetry potentials, which necessitates regular standardization; it is very sensitive to previous treatment and should be well washed after use.

Glass electrodes in combination with calomel reference electrodes have been used to follow titrations in non-aqueous solvents: for example, the titration of perchloric acid with sodium salts of organic acids in glacial acetic acid. See Ives and Janz (1961).

Further reading
R. G. Bates, *Determination of pH, Theory and Practice*, John Wiley, New York (1973)

Glucose electrode. The first immobilized enzyme electrode was designed by Clark and Lyons (1962) for the determination of glucose (*see* **Biosensor**). It uses immobilized glucose oxidase on an amperometric sensor. In the presence of glucose, oxygen is consumed

$$\text{glucose} + O_2 + H_2O \xrightarrow{\text{glucose oxidase}} H_2O_2 + \text{gluconic acid}$$

Changes in the partial pressure of oxygen are detected using an amperometric oxygen electrode

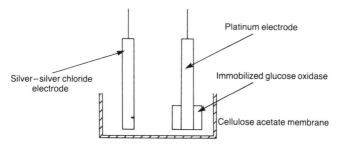

Figure 75. Schematic diagram of a potentiometric enzyme electrode for monitoring glucose levels

$$O_2 + 2H_2O + 4e \rightarrow 4OH^-$$

Alternatively, the production of hydrogen peroxide may be monitored amperometrically

$$H_2O_2 \rightarrow O_2 + 2H^+ + 2e$$

A potentiometric glucose electrode is based on an iodide ion-selective electrode. Glucose oxidase and peroxidase (2:1) are immobilized in a layer over the iodide electrode. The hydrogen peroxide liberated by the glucose oxidase reaction oxidizes iodide ions (added to the sample solution), and the local changes in the iodide concentration are detected

$$H_2O_2 + 2I^- + 2H^+ \rightarrow I_2 + 2H_2O$$

A glass electrode that carries a thin film of glucose oxidase can be used to detect pH changes resulting from the generation of gluconic acid.

Glucose oxidase is highly specific for β-D-glucose. Interference has been observed by the following: cellibiose, maltose, 2-deoxyglucose and such reducing substances as uric acid, tyrosine, ascorbic acid and iron (II).

A recent development is the immobilization of glucose oxidase, with or without catalase, on platinum (*see* Fig. 75). In the presence of glucose, the enzymic reaction generates low levels of hydrogen peroxide, which in turn interacts with the oxide layer on the platinum to generate a potential. The Nernst equation is valid over the range 100–4000 mg dm^{-3} (normal blood level, 650–1100 mg dm^{-3}).

Goüy Layer. *See* **Electrical double layer.**

Grotthus theory. *See* **Molar ionic conductivity.**

H

Half cell. If a metal is placed in a solution of its salt, a 'couple' or 'half cell' is formed; the oxidized form (ions in solution) and the reduced form (metal) are in contact with each other. At the surface, separation of charge occurs and an **interfacial potential** difference—the **electrode potential**—is established.

In some metals (e.g., zinc), the atoms lose electrons and pass into solution

$$M \rightarrow M^{z+} + ze$$

This process results in the accumulation of liberated electrons in the metal, which therefore becomes negatively charged with respect to the solution. On the other hand, with some metals (e.g., copper) the reverse process occurs

$$M^{z+} + ze \rightarrow M$$

leading to a deficit of electrons in the metal which thereby becomes positively charged.

In general, the number of ions deposited or released in this way is very small and the tendencies for the two processes to occur are soon reduced to zero by the potential differences so established. As a negative potential develops on M, the rate of its ionization decreases, and on the other hand, the rate at which ions are discharged increases until the equilibrium

$$M \rightleftharpoons M^{z+} + ze$$

is established when the two rates are equal. The final potential difference—the electrode potential—which is established depends on the activity of the ions in solution and the temperature. If two such half cells have different electrode (reduction) potentials then when the two electrolyte solutions are suitably connected (e.g., by a **salt bridge**, porous plate or direct contact with minimal mixing of the solutions), an e.m.f. is produced as a result of an overall chemical reaction arising from the oxidation and reduction reactions at the two electrodes.

A reversible half cell (electrode) has the following characteristics when combined with a hydrogen electrode in a cell: (1) there is no reaction before the cell is completed; (2) at the equilibrium potential the electrode reaction will proceed in a forward or reverse direction when the external e.m.f. is decreased or increased by an infinitesimally small amount and (3) the electrode reaction proceeds to such a small extent that the activity of the

Half cell

reactive ion in solution is virtually unchanged and hence the electrode potential remains constant over a reasonable period of time. All electrode equilibria involve two opposing reactions

$$ox + ne \rightleftharpoons red$$

and for such an electrode, the reduction potential relative to the standard **hydrogen electrode** is given by the **Nernst equation**

$$E = E^\ominus + \frac{RT}{nF} \ln \frac{a_{ox}}{a_{red}}$$

In representing half cells a vertical line indicates the junction of the two phases, for example

$$Zn^{2+} \mid Zn \qquad H^+ \mid H_2, Pt.$$

The following types of half cells are recognized:

1) Solid element in contact with a solution of its ions (*see* **Amalgam electrode; Metal electrode**).
2) **Redox electrode system**, an inert metal immersed in a solution containing ions in different states of oxidation, for example

$$Fe^{3+}, Fe^{2+} \mid Pt$$

(*see also* **Biological oxidation–reduction systems**).

3) Inert electrode, neutral solutes in different states of oxidation, for example

$$\text{quinone, hydroquinone} \mid Pt$$

(*see* **Quinhydrone electrode**).

4) **Gas electrode**, inert electrode, gas in equilibrium with its ions in solution, for example

$$H^+ \mid H_2, Pt \qquad Pt \mid Cl_2, Cl^-$$

5) Metal in contact with a solution saturated with a sparingly soluble salt of the metal and a salt with a common anion, for example

$$AgCl(s), Ag \mid Cl^-$$

(*see* **Antimony oxide electrode; Calomel electrode; Mercury–mercury (II) oxide electrode; Silver–silver chloride electrode**). The silver–silver chloride and calomel electrodes are commonly used as reference electrodes in potentiometric titrations (*see* **Electrometric titrations**) and for the measurement of standard electrode potentials.

6) Cationic-responsive glass electrodes (*see* **Glass electrode**).
7) Cationic-responsive electrode. This consists of a metal, one of its insoluble salts and another insoluble salt with a common anion but different cation

154 Half-wave potential

immersed in a solution containing the common cation: for example

$$Ca^{2+} \mid CaC_2O_4(s) \mid PbC_2O_4(s) \mid Pb$$

behaves as a reversible calcium electrode with an electrode potential given by

$$E_{Ca2+,Ca} = E' + \frac{RT}{2F} \ln a_{Ca2+}$$

where E' is not a true standard electrode potential but includes the solubility products of the sparingly soluble salts. See Ives and Janz (1961).

Half-wave potential. *See* **Polarography.**

Hall–Héroult process. *See* **Aluminium electrometallurgy.**

Halogen electrode reaction. The electrochemical oxidation of the halide ion, and reduction of the halogen molecule in solution or melt

$$2X^- \rightleftharpoons X_2 + 2e$$

where X is fluorine, chlorine, bromine or iodine.

The production of the halogens, in particular chlorine, is of great industrial importance. Chlorine is evolved by a mechanism in which the discharge of chloride ion is rate-determining

$$Cl^- \xrightarrow{slow} Cl_{ads} + e$$
$$2Cl_{ads} \longrightarrow Cl_2$$

The slow step in the evolution of bromine from aqueous bromide solution is

$$Br^- + Br_{ads} \rightarrow Br_2 + e$$

Iodine follows a similar reaction path to that of chlorine, although the reaction is complicated by the equilibrium with triiode (I_3^-)

$$I^- + I_2 \rightleftharpoons I_3^-$$

See also **Electrolysis of brine.**

Helmholtz layer. *See* **Electrical double layer.**

Henderson equation. *See* **Buffer solution; Dissociation constant; Electrometric titrations.**

Heyrovsky reaction. *See* **Hydrogen electrode reactions.**

Hittorf method. *See* **Transport number.**

Hydration of ions. *See* **Molar ionic conductivity.**

Hydrocarbon–air cell. *See* **Fuel cell.**

Hydrogen electrode. When platinum coated with platinum black (*see* **Platinum and gold electrodes**) is saturated with hydrogen gas and immersed in a solution containing hydrogen ions, it behaves like a metal electrode

$$\tfrac{1}{2}H_2(g) \rightleftharpoons H^+ + e$$

The **electrode potential** is given by

$$E_{H+,H2} = E^{\ominus}_{H+,H2} + \frac{RT}{F} \ln \frac{a_{H+}}{(p_{H2}/101\,325)^{1/2}}$$

where p_{H2} is the partial pressure of hydrogen in N m^{-2}. The standard hydrogen electrode potential, in which $p_{H2} = 1$ atmosphere ($101\,325$ N m^{-2}) and $a_{H+} = 1$, is by convention the arbitrary zero of electrode potential. Thus $E^{\ominus}_{H+,H2} = 0$ and hence

$$E_{H+,H2} = \frac{RT}{F} \ln \frac{a_{H+}}{(p_{H2}/101\,325)^{1/2}}$$

The simple form of the hydrogen electrode (*see* Figure 76) consists of a piece of platinized platinum fused into glass tubing which carries the electrical connection. Hydrogen, free from oxygen, is first passed through water and a sample of the solution under test in the cell, to avoid altering the concentration of the solution in the cell, and then into the test solution in the cell. In operation, a steady stream of hydrogen is bubbled through the solution until

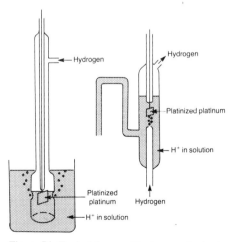

Figure 76. Typical forms of hydrogen electrode

156 Hydrogen electrode reactions

the cell (comprising the hydrogen electrode and a reference electrode) assumes a constant e.m.f. This may take 10–15 minutes; thereafter the e.m.f. is independent of the rate of bubbling.

The hydrogen electrode is capable of a high degree of accuracy, giving reproducible results over the complete pH range 0–14; there is no salt error (i.e. no apparent shift of pH caused by variation of the ionic strength). However, it cannot be used in the presence of air, oxygen or oxidizing agents and reducing agents; the platinized surface is readily poisoned by alkaloids, cyanides, arsenic and antimony compounds, and by colloids which are adsorbed on the surface.

The hydrogen electrode is widely used as a standard electrode in non-aqueous solvents, and its potential, relative to the standard electrode in aqueous solution, has been established in many solvents (*see* Appendix, Table 4). In liquid ammonia, the potential of the hydrogen electrode is very reproducible although the absolute value is sensitive to the partial pressure of hydrogen which is difficult to control near the boiling point of ammonia. See Ives and Janz (1961).

Hydrogen electrode reactions
Cathodic reaction. Hydrogen is electrochemically generated by reactions having the overall stoichiometry

$$2H^+ + 2e \rightarrow H_2 \qquad \text{in acidic solution} \qquad (a)$$

and

$$2H_2O + 2e \rightarrow H_2 + 2OH^- \qquad \text{in alkaline solution} \qquad (b)$$

There are two reaction pathways. Both involve the initial electrochemical discharge of a proton (the Volmer reaction) on the metal electrode surface M.

$$M + H^+ + e \rightarrow MH_{ads} \qquad (c)$$

This may be followed by either a chemical step in which two adsorbed hydrogen atoms react to give a molecule (the Tafel reaction)

$$2MH_{ads} \rightarrow 2M + H_2 \qquad (d)$$

or a second electrodic step (the Heyrovsky reaction)

$$MH_{ads} + H^+ + e \rightarrow M + H_2 \qquad (e)$$

The actual mechanism under any given conditions depends, in the main, on the nature of the electrode surface. Determination of the mechanism is by measurement of various electrochemical parameters (Tafel slope, effect of pH on exchange current density, etc.; *see* **Electrode reaction mechanisms**). The theoretical values for the combinations of mechanisms with different rate-determining steps are given in Table 18. Although many methods give

Table 18
Theoretical kinetic parameters for possible mechanisms of the hydrogen evolution reaction, assuming $\beta \doteq 0.5$

Mechanism	Tafel slope $(\partial \eta/\partial \ln i)$	$\left(\dfrac{\partial \eta}{\partial \ln a_{H+}}\right)_i$	$\left(\dfrac{\partial \eta}{\partial \ln p_{H2}}\right)_i$	$\left(\dfrac{\partial \eta}{\partial \ln \theta}\right)_i$	ν^a	Tritium separation factor
Slow discharge, fast recombination (e.g., mercury)	$\dfrac{2RT}{F}$	$\dfrac{RT}{F}$	$\dfrac{RT}{2F}$	0	2	5
Fast discharge, slow recombination (e.g., platinum)	$\dfrac{RT}{2F}$	0	$\dfrac{RT}{2F}$	$-\dfrac{RT^b}{F}$ 0^c	1	11
Fast discharge, slow electrodic discharge (e.g., copper)	$\dfrac{RT^b}{1.5F}$ $\dfrac{2RT^c}{F}$	$\dfrac{RT}{F}$	$\dfrac{RT}{2F}$	$-\dfrac{RT^b}{F}$ 0^c	1	23

a The stoichiometric number.
b If the coverage of adsorbed hydrogen atoms (θ) is low ($\theta \to 0$).
c $\theta \to 1$.

similar predictions, the hydrogen–tritium separation factor is probably the most unambiguous mechanistic indicator.

Whether the rate-determining step involves the making or breaking of a metal–hydrogen bond may be inferred from the position of the metal in a series which relates the rate of hydrogen evolution to the metal–hydrogen bond strength. If the rate-determining step is bond-breaking, an increase in metal–hydrogen bond strength *decreases* the rate, whereas the converse is true when bond making is the rate-determining step. Figure 44 shows that only for the most inactive metals (e.g., lead, mercury, cadmium and thallium) is bond making (reaction (c)) rate-determining. For these metals, the slow electrochemical discharge of a proton is followed by fast recombination of adsorbed hydrogen atoms (reaction (d)). For other metals (e.g., nickel, copper, gold and silver), electrochemical discharge of a proton is fast, with a slow electrodic discharge or, for more active metals (e.g., platinum and rhodium), a slow chemical recombination. Very high surface area platinum can have such a great rate of hydrogen evolution that the process is limited by diffusion of the generated hydrogen molecules away from the electrode surface.

Anodic reaction. By the principle of microscopic reversibility, the mechanism of hydrogen oxidation must be the reverse of the corresponding evolution reaction at the reversible potential. A complication arises for many non-noble metals in acidic solution, in that anodic reactions of the metal (dissolution, oxide or hydroxide formation) are also thermodynamically feasible in the potential region of hydrogen oxidation. The mechanism on platinum is thought to be the reverse of evolution.

Oxidation of hydrogen at platinum-doped tungsten trioxide shows evidence of the participation of the hydrogen tungsten bronze H_xWO_3

$$2Pt + H_2 \rightarrow 2PtH_{ads}$$
$$xPtH_{ads} + WO_3 \rightarrow xPt + H_xWO_3$$

Oxidation of the tungsten bronze adds to the anodic current leading to a so-called 'synergistic' effect or, more exactly, to 'spill over' catalysis (referring to the spill over of hydrogen atoms from platinum to tungsten trioxide). *See also* **Fuel cell**.

Hydrogen embrittlement. The loss of strength of a metal after exposure of the surface to hydrogen. This may come about from the use of the metal to evolve hydrogen electrochemically. A more insidious occurrence is in **corrosion** when parts of a metal structure become sites for a reduction reaction, which in certain circumstances may be the evolution of hydrogen. Hydrogen embrittlement occurs only with hydrogen atoms which diffuse into the metal lattice. These, by themselves, will distort and weaken the lattice, but the presence of small voids or cracks in the structure causes molecular hydrogen to accumulate and, when a sufficiently high pressure has built up, crack the metal further.

Cathodic protection of a metal may, if too great a potential is applied, generate hydrogen and cause embrittlement. A possible solution to the problem lies in speeding up the recombination of adsorbed hydrogen atoms (*see* **Hydrogen electrode reactions**). A coat of a catalyst, such as $NiCo_2S_4$, on a metal surface can lower the surface concentration of hydrogen atoms to a level at which diffusion into the lattice is substantially reduced. *See also* **Stress corrosion cracking.** See Bockris and Reddy (1973).

I

Ilkovic equation. *See* **Polarography**.

Impedance (Z). The reactance of a capacitor or an inductance has the same dimensions as resistance, but it differs from resistance in being dependent on the frequency of the ac supply. For a capacitor

$$X_C = \frac{1}{2\pi\nu C} = \frac{1}{\omega C}$$

where ν is the frequency and ω the angular frequency of the ac supply. The reactance of an inductance is

$$X_L = \omega L$$

where L is the inductance of the circuit.

If a reactance X and resistance R are joined in series the effective combined resistance of the combination Z is called the impedance. The magnitude of the impedance is given by

$$Z = (R^2 + X^2)^{1/2} \tag{83}$$

Since the reactances of capacitors and inductances are 180° out of phase, the combined reactance for such a series arrangement is

$$X = \frac{1}{\omega C} - \omega L$$

whence Eq. (83) becomes

$$Z = \left[R^2 + \left(\frac{1}{\omega C} - \omega L\right)^2\right]^{1/2}$$

The phase angle θ by which the current leads the voltage is given by

$$\tan \theta = \frac{X}{R}$$

The current produced by an e.m.f. E in a simple reactive circuit is given by

$$I = \frac{E}{Z} = \frac{E}{(R^2 + X^2)^{1/2}}$$

The reciprocal of the impedance is known as the admittance.

Indicator electrode. The definition of suitable indicator electrodes forms the basis of both theoretical and practical aspects of e.m.f. measurements: for example, in obtaining basic thermodynamic data, elucidation of electrode processes and *in situ* analyses. The three that meet the requirements most adequately are the **hydrogen electrode**, the **silver–silver chloride electrode** and the **calomel electrode**. The only electrode to which the term reference electrode may be rigorously applied is the hydrogen electrode. Of the secondary electrodes, a distinction is drawn between those for which the standard potential can be expressed in terms of strictly thermodynamic quantities and those which are less easily explained on purely thermodynamic principles (e.g., the **glass electrode**).

Some applications of different electrodes are listed in Table 19.

Table 19

Electrode	Suitability[a] Aqueous solutions	Organic solvents	Biological systems	Special applications and reservations
Hydrogen	4	2	1	Electrode standardization
Silver–silver halide	4	3	3	Secondary reference electrode, thermodynamic studies
Calomel	4	2	2	Thermodynamic studies (contaminates solution)
Glass	4	2	4	pH measurements, potentiometric titrations
Quinhydrone	4	1	1	Potentiometric studies (contaminates solution)
Metal–metal oxide	3	1	1	Best used in alkaline range
Oxygen	4	1	4	Measurements of oxygen tension

[a] 1, inapplicable or insufficient data information available; 2, occasionally suitable; 3, applicable to selected systems; 4, generally applicable.

Indicators.

Acid–base indicators. The pH of the exact equivalence point of an acid–base titration depends on the relative strengths of the acid and the base (*see* **Electrometric titrations**). The indicator must exhibit its colour change as close as possible to the pH of the equivalence point.

Acid–base indicators are generally weak acids which possess different colours according to the pH of the solution. Classically

$$HIn \rightleftharpoons H^+ + In^-$$

for which the apparent ionization constant (neglecting activity coefficients) is

162 Indicators

$$K_{In} = \frac{c_{H^+} c_{In^-}}{c_{HIn}}$$

or

$$\mathrm{pH} = pK_{In} + \log \frac{c_{In^-}}{c_{HIn}}$$

Thus, when an indicator is added to a solution of given pH, the equilibrium adjusts itself until these equations are obeyed. Since the two forms of the indicator—HIn and In$^-$—have different colours, the colour of the indicator and, hence, of the solution is therefore adjusted. From the ratio of the intensities of the two colours the pH of the solution may be determined. Because of the difficulty of detecting a small intensity of one colour in the presence of another, the useful range of an indicator is limited to

$$\mathrm{pH} = pK_{In} \pm 1$$

This enables the correct choice of indicator for a given titration (*see* Appendix, Table 8).

The ionization constant of the indicator may be determined by methods used for the dissociation constant of an acid, especially the spectrophotometric method.

Oxidation–reduction indicators. When added to a large amount of redox system, the indicator adopts the potential of the system and the ratio of the intensity of the colour of the oxidized and reduced forms takes up the appropriate value. From a knowledge of this ratio, the potential of the system can be determined

$$E = E_{In}^{\ominus} + \frac{RT}{nF} \ln \frac{c_{ox}}{c_{red}}$$

The indicator used to determine the equivalence point in an oxidation–reduction titration should have a standard potential midway between the redox potentials of the two reagents in the titration. These indicators, generally organic dyestuffs with different colours in the oxidized and reduced forms, have a limited range of use;

$$E = E_{In}^{\ominus} \pm 0.06$$

For a redox titration to be quantitative, the standard redox potentials of the two systems must differ by at least 0.3 V; the rapid change of potential at the end-point should thus cover a range of at least 0.12 V. It is possible to achieve a very sharp end-point with a redox indicator provided that its working range is covered by the change of potential at the end-point (*see* Appendix, Tables 9 and 10). To overcome some of the limitations of visual indicators (e.g., in highly coloured solutions) electrometric titrations have been developed to locate the end-point. See James (1976); James and Prichard (1977).

Indifferent electrolyte. An added constituent to an electrolyte solution which takes no part in the electrode processes under study. Its functions are to reduce the resistance of the solution and, by carrying almost all the current, to ensure that the electroactive constituents (i.e. those taking part in the electrode reactions) reach the electrode surfaces by diffusion and not by electrolytic transport.

Inner potential. *See* **Interfacial potential.**

Interfacial potential (Galvani potential) (ϕ). For a phase, the interfacial potential is the work necessary to move a unit charge from infinity to the interior of the phase. For the metal ion–metal electrode $M^{z+} \mid M$

$$M^{z+} + ze \rightleftharpoons M$$

The phase boundary potential $\Delta\phi_{\alpha\beta}$ (Eq. (57)) becomes

$$\Delta\phi = \phi_M - \phi_{soln} = \frac{\mu^\ominus_{soln} - \mu_M}{zF} + \frac{RT}{zF} \ln a_+$$

$$= \Delta\phi^\ominus + \frac{RT}{zF} \ln a_+$$

$\mu^\ominus_{soln}$ and μ_M are constant and independent of concentration of M^{z+} in solution. Thus the potential of a metal becomes more positive the greater the concentration of M^{z+} in solution.

For an electrode reversible with respect to the anion

$$\phi_{solid} - \phi_{soln} = \Delta\phi^\ominus - \frac{RT}{zF} \ln a_-$$

The interfacial potential is the sum of the **surface potential** and the **outer potential**.

$$\phi = \psi + \chi$$

The measured e.m.f. of a cell is the difference in the Galvani potentials of two identical metallic terminals attached to the electrodes. Although the e.m.f. is the only measurable quantity, it is actually composed of two or more Galvani potential differences arising at the individual phase boundaries. *See also* **Electrochemical potential.**

Interruptor technique. The potential measured between the working and reference electrodes in a **three-electrode system** includes a component due to

Iodine coulometer

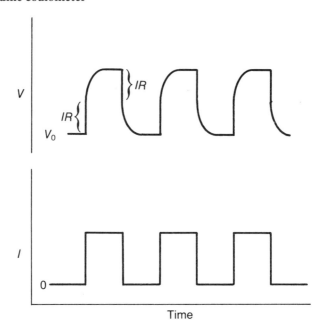

Figure 77. Variation of potential with time as the current is interrupted at 50 Hz

the resistance of the solution, the so-called *IR* drop. The use of a **Luggin capillary** minimizes the *IR* drop, but at high currents, the measured overpotential must be corrected for this component. The interruptor technique relies on the fact that ohmic potentials decay much faster than activation or concentration overpotentials (*see* **Overpotential**) when the current falls to zero. In a galvanostatic circuit, the current is interrupted at 50 Hz (60 Hz in the USA) by means of a mercury relay. The potential between working and reference electrodes is monitored on an oscilloscope and has the form shown in Figure 77. The sudden rise and fall in the potential is due to ohmic losses and should be subtracted from the total potential change when, for example, a Tafel plot is made (*see* **Tafel equation**).

Iodine coulometer. *See* **Coulometer.**

Ion association. *See* **Ion pair.**

Ionic atmosphere. *See* **Conductance equations; Electrical double layer.**

Ionic melts. *See* **Fused salts.**

Ionic mobility. *See* **Molar ionic conductivity.**

Ionic product of water

Ionic product of water (K_w; units: $mol^2\ dm^{-6}$). Pure water has a small electrical **conductivity** due to self-ionization

$$H_2O + H_2O \rightleftharpoons H_3O^+ + OH^-$$

This process of proton transfer is known as autoprotolysis (*see* **Acids and bases**). For this equilibrium, since $a_{H_2O} = 1$

$$K_w = a_{H_3O^+} a_{OH^-}$$
$$pK_w = -\log K_w$$

The ionic product principle is valid in any aqueous solution irrespective of its pH. For all acids and conjugate bases in aqueous solution, the product of the acidic and basic dissociation constants equals K_w.

The value of K_w has been determined from accurate conductivity measurements on conductivity water. It may also be determined from e.m.f. studies on cells without liquid junctions: for example

$$\ominus \quad \begin{array}{c} Pt, H_2(g) \\ (101\,325\ N\ m^{-2}) \end{array} \left| \begin{array}{cc} KOH, & KCl \\ (m_1) & (m_2) \end{array} \right| AgCl, Ag \quad \oplus$$

for which

$$\frac{F(E - E^\ominus_{Ag,AgCl,Cl^-})}{2.303 RT} + \log \frac{m_2}{m_1} = -\log K_w - \log \frac{\gamma_{Cl^-}}{\gamma_{OH^-}} \quad (84)$$

The e.m.f. (E) of the cell is measured for known values of the molality of potassium chloride and potassium hydroxide (m_1 and m_2, respectively), the graph of the left-hand side of Eq. (84) against the square root of the ionic strength is linear and of intercept $-\log K_w$.

The ionic product of water varies with temperature (*see* Table 20). Since at 25°C, $K_w \approx 10^{-14}\ mol^2\ dm^{-6}$, the concentrations (activities) of the hydrogen and hydroxyl ions are equal to $10^{-7}\ mol\ dm^{-3}$; thus the pH of pure water is 7.0.

Table 20
Variation of the ionic product of water with temperature

T/°C	$10^{14}\ K_w/mol^2\ dm^{-6}$
0	0.116
10	0.281
18	0.590
25	1.008
40	2.919
50	5.660
60	9.614

Ionic strength

Ionic strength (I; units: mol dm^{-3}, mol m^{-3}, mol kg^{-1}). A characteristic of an electrolyte solution defined by

$$I = \tfrac{1}{2} \sum_i m_i z_i^2$$

where the summation extends over all ions in solution. In terms of volume concentration

$$I = \tfrac{1}{2} \sum_i c_i z_i^2$$

I is related to the concentration (m or c) by the relation $I = km$ or kc; the value of k depends on the valences of the ions in solution. Thus for a solution of sodium sulphate of molality m

$$\begin{aligned}I/\text{mol kg}^{-1} &= \tfrac{1}{2}(m_{\text{Na}+}z_{\text{Na}+}^2 + m_{\text{SO4,2}-} z_{\text{SO4,2}-}^2) \\ &= \tfrac{1}{2}(2m + 4m) = 3m\end{aligned}$$

Values of k for various valence type electrolytes are presented in Table 21. Non-ideal behaviour in electrolytes solutions is primarily due to the total concentration of ions present, rather than to their chemical nature; the ionic strength provides a method for the expression of the total ionic concentrations for all types of electrolyte. It is thus possible to compare the effects of different electrolytes (e.g., on the rates of reaction in solution).

Table 21

	M$^+$	M^{2+}	M^{3+}	M^{4+}
A$^-$	1	3	6	10
A^{2-}	3	4	15	12
A^{3-}	6	15	9	42
A^{4-}	10	12	42	16

Ion pair. For many electrolytes (particularly those of higher valence type), the experimental values of molar conductivity are much lower than the theoretical values predicted by Onsager's equation (*see* **Conductance equations**). It was suggested by Bjerrum that, under certain conditions, the oppositely charged ions of an electrolyte associate to form ion pairs and even triple and quadruple ions. Such association, which leads to a smaller number of particles in solution with a lower charge (often zero) than the unassociated ions, will result in a decrease in the molar conductivity and other properties dependent upon the number and charge of particles in solution. Such aggregates of ions exist to an appreciable extent in relatively concentrated solutions and particularly in fused salts, but are easily dispersed on dilution; thus the molar conductivity increases with decrease in concentration.

The basic theory is that the Debye–Hückel theory is valid so long as the oppositely charged ions of the electrolyte are separated by a critical distance q greater than the minimum value given by

$$q = \frac{|z_+||z_-|e^2}{8\pi\varepsilon_0\varepsilon_r kT}$$

At this distance, the mutual electrical potential energy between two ions is $2kT$, that is the energy necessary to separate the ion pair is comparable to their thermal energy. When the separation is less than q, ion pairing is considered to have occurred. For electrolytes in water ($\varepsilon_r = 80.36$) at 298 K, $q = 0.349|z_+||z_-|$ nm, while in dioxan ($\varepsilon_r = 2.24$) $q = 12.5|z_+||z_-|$ nm. Should the sum of the ionic radii be less than this, ion pair formation will be favoured (e.g., small highly charged ions in solvents of low relative permittivity and ions of high valence type in aqueous solution). For 1:1 electrolytes, there is virtually no association, but complete dissociation, in solvents with relative permittivity greater than 42; this has been verified by conductance measurements.

The equilibrium between free ions and ion pairs can be formulated in the same way as for a weak electrolyte

$$K_D = \frac{\gamma_+\gamma_-\alpha^2 m}{(1-\alpha)}$$

where K_D is the dissociation constant and α is the fraction of electrolyte present as free ions. The extent of ion pairing varies from an amount too small to measure accurately in aqueous solutions of potassium nitrate to 75 percent ion pairing for manganese (II) oxalate at a concentration of 5×10^{-3} mol dm^{-3}.

The Bjerrum theory accounts for the results satisfactorily in some cases, but it is clear that in others additional factors must be taken into account. One of these is the hydration of the free ions and the ion pairs.

For electrolytic dissociation arising by a solvolytic reaction with water or an alcohol

$$HCl + H_2O \rightleftharpoons H_3O^+(aq) + Cl^-(aq)$$

ion pair formation between already dissociated molecules must be distinguished from incomplete dissociation of the molecule itself. For acids and bases which ionize by reaction with the solvent, it is possible to distinguish the ion pairs from the undissociated molecules by Raman spectroscopy. Dissociation by solvolysis probably occurs by an ion pair intermediate

$$HCl + H_2O \rightleftharpoons (H_3O^+Cl^-)_{pair,aq} \rightleftharpoons H_3O^+(aq) + Cl^-(aq)$$

168   Ion pair

In non-aqueous media, the nature of the ion pair involved in dissociation reactions may be elucidated by Raman, ultraviolet and nuclear magnetic resonance spectroscopic measurements.

Three types of ion pairs may be distinguished (*see* Figure 78).

1) Solvent-separated ion pairs in which the solvation shells of the anions and cations are more or less retained in their undisturbed configuration as the ions become paired. Longer range solvation is probably changed.

2) Solvent-shared ion pairs in which the pairing of two solvated ions results in the loss of some associated solvent by each of them but with at least one solvent molecule shared between the ions in their respective solvation shells forming a new ion pair solvate configuration with a new minimum free energy.

3) Contact ion pairs in which most of the primary solvation is lost by both ions, especially in the direction between the centres of the ions, so that the ions are in contact.

Ion pairs of type 1 are probably more common especially in highly polar solvents, while type 3 arises in weakly polar solvents and/or with pairs of ions having large ionic radii (e.g., tetra-iso-amylammonium picrate, tetra-n-butylammonium tetra-phenylboride). In many cases, two or three types of ion pairs may be in equilibrium (e.g., in magnesium sulphate) and their presence in solution can be distinguished by their rate constants for formation or by their dissociation constants. In aqueous solutions, ion pairs of types 1 and 3 are probably not involved.

On dissociation there is an appreciable increase in **electrostriction**, which is in accord with the expected greater coulombic influence of the fields of the separate ions on the water than that of the paired ions.

The formula of the ion pair and the dissociation constant of an incompletely dissociated electrolyte can be obtained from spectrophotometric measurements in the ultraviolet region of the spectrum, provided that the two species involved absorb at different wavelengths: for example, the dissociation constant for the ion pair Pb NO_3^+ is 0.62 mol dm^{-3}. Values of K_D for copper sulphate obtained from spectroscopic, conductivity and cryoscopic measurements—3.5×10^{-3}, 3.9×10^{-3} and 3.3×10^{-3} mol dm^{-3}, respectively—are in good agreement. See Davies (1962); Robinson and Stokes (1970).

Further reading

B. E. Conway, Ionic interaction and activity behaviour of electrolyte solutions, in *Comprehensive Treatise on Electrochemistry*, vol. 5, J. O'M. Bockris *et al.* (ed.), Plenum Press, New York, chapter 2 (1982)

T. H. Lilly, Raman spectroscopy of aqueous electrolyte solutions, in *Water, a Comprehensive Treatise*, vol. 3, F. Franks (ed.), Plenum Press, New York, p. 265 (1973)

D. R. Rosseinsky, Interactions involving aquo-ions, *Ann. Rep. Chem. Soc.*, **A68**, 82 (1971)

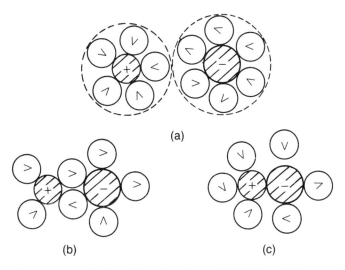

Figure 78. Possible structures of solvated ion pairs: (a) type 1, hydration shell contact; (b) type 2, shared hydration shell; (c) type 3, ion contact type

Ion-selective electrode. A membrane electrode that responds selectively to one (or several) ionic species in the presence of other ions. The word 'membrane' is used in its widest sense denoting a thin layer of electrically conducting material separating two solutions across which a potential develops. The first ion-selective electrode was the hydrogen ion-responsive **glass electrode**.

Ion-selective electrodes are classified according to the IUPAC suggestion (*see* Table 22).

Table 22
Typical ion-selective electrodes

Solid state	Heterogeneous	Liquid ion exchanger	Glass
F^- (LaF_3)	F^- (LaF_3)	Cl^-	H^+
Cl^- (AgCl)	Cl^- (AgCl)	ClO_4^-	Na^+
Br^- (AgBr)	Br^- (AgBr)	NO_3^-	K^+
I^- (AgI)	I^- (AgI)	Ca^{2+}	NH_4^+
S^{2-} (Ag_2S)	S^{2-} (Ag_2S)	Cu^{2+}	Ag^+
Ag^+ (AgCl, AgBr, AgI)	Ag^+ (AgCl, AgBr, AgI)	Pb^{2+}	Li^+
Cu^{2+} (Ag_2S + CuS)	SO_4^{2-} ($BaSO_4$)	BF_4^-	Ca^{2+}
Pb^{2+} (Ag_2S + PbS)	PO_4^{3-} ($BiPO_4$)	K^+	
Cd^{2+} (Ag_2S + CdS)			
CN^- (AgI, Ag_2S)			

170 Ion-selective electrode

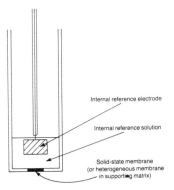

Figure 79. Schematic representation of a homogeneous/heterogeneous membrane electrode

Primary electrode.
1) Crystalline.
a) Homogeneous membrane electrode: A solid-state electrode based on a single crystal or compacted disc of an insoluble salt (e.g., AgX) or a mixture (e.g., silver iodide/silver sulphide) sealed on the end of a tube containing a reference solution and electrode, usually a silver–silver chloride electrode (*see* Figure 79). A crystal of lanthanum fluoride sealed in a tube permits the direct determination of fluoride ions in solution in the range 10^{-1}–10^{-6} mol dm^{-3}.
b) Heterogeneous membrane electrode in which the active material (e.g., precipitates of insoluble metal salts) is dispersed in an inert matrix to give suitable mechanical properties (e.g., silver iodide in silicone rubber). The prepared membrane is cemented onto the end of a tube containing an internal reference solution and electrode (*see* Figure 79). Silver halide and sulphide electrodes respond to the silver ion (10^{-1}–10^{-5} mol dm^{-3}) and the corresponding anion (10^{-1}–10^{-6} mol dm^{-3}).
2) Non-crystalline, with or without support.
a) Rigid matrix glass electrodes which are responsive to potassium, sodium, ammonium or silver ions are obtained by varying the composition of the glass. Selectivity is not good when other metal ions are present. The principal applications of such cation-responsive electrodes are in water analysis and in clinical biochemistry.
b) Liquid ion-exchanger electrodes in which the ion of interest is incorporated in a large organic molecule with low water solubility. The organic material dissolved in an organic solvent is separated from the aqueous solution under study by a porous membrane (e.g., cellulose acetate) holding the liquid ion exchanger (*see* Figure 80). Anion exchangers are either long-chain alkyl ammonium salts or salts of a non-labile metal complex of the type ML_3X_2,

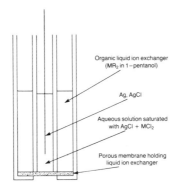

Figure 80. Cross-section of a liquid membrane electrode sensitive to M^{2+}

where, for example, L for the perchlorate and nitrate electrodes is a substituted 1,10-phenanthroline.

Cationic exchangers usually consist of a long alkyl chain compound with low water solubility which forms a stable compound with the cation under study. The calcium electrode, using calcium bis(di-n-decyl)phosphate, $[(C_{10}H_{21})_2PO_2]_2Ca$, dissolved in di-n-octylphenylphosphonate, shows Nernstian behaviour (29.4 mV/pCa) over the range $1-10^{-5}$ mol dm^{-3}. The performance is independent of pH in the range 5.5–11. At lower pH values, hydrogen ion exchange with calcium ions on the exchanger occurs to a significant extent; the electrode then becomes pH- as well as pCa-dependent.

Cation-responsive electrodes based on neutral charge carrier ligands are widely used; the antibiotic valinomycin and the macrotetrolide antibiotics are the most attractive complexing agents for the potassium and ammonium ions. The antibiotic ionophores satisfy the criteria for ion carrier function, with both polar and non-polar groups which can adopt a stable configuration providing a cavity surrounded by polar groups with an exterior hydrophobic shell of non-polar groups. The cation with the best 'fit' within the cavity provided is selectively complexed. Crown ethers with nine- to 60-membered rings and containing three to 20 ether oxygen atoms, which form stable complexes with alkali and alkaline earth cations in non-polar solvents, have an advantage over the antibiotics in that they can be tailored to achieve desired specificity for a particular cation.

PVC–matrix membrane ion-selective electrodes are an extension of liquid-exchanger electrodes in which the liquid ion exchanger is incorporated in PVC with the aid of a solvent, which on evaporation leaves a flexible membrane with the ion-exchanger components trapped in the matrix. This membrane is then cemented to the end of a tube containing a reference solution and electrode (compare Figure 79). Coated wire electrodes can be prepared by dipping a platinum wire or graphite rod into a solution of PVC/sensor/ mediator in a solvent and allowing the solvent to evaporate. The 'Selectrode', devised by Ruzicka, consists of a porous graphite rod made hydrophobic by treating with Teflon; the active material is simply rubbed on the surface or applied by hot dipping.

Sensitized electrodes.
1) **Gas-sensing membrane probe**
2) **Biosensor**, biocatalytic membrane electrode or enzyme electrode

In operation these electrodes can be considered as being part of a **concentration cell**, in which the active material acts as a 'bridging electrode'. Thus the lanthanum fluoride electrode, which is selective for fluoride ions, can be considered as a double cell

$$\text{Ag,AgCl} \;\vert\; \begin{array}{c}\text{NaF,NaCl}\\(m_1)\end{array} \;\vert\; \text{LaF}_3 \;\vert\; \begin{array}{c}\text{NaF,NaCl}\\(m_2)\end{array} \;\vert\; \text{AgCl,Ag}$$

$\leftarrow$ ion-selective electrode $\rightarrow$ $\;\;$ test solution

in which the lanthanum fluoride crystal replaces the two lanthanum–lanthanum fluoride electrodes. Such an electrode cannot be constructed because lanthanum metal oxidizes readily in air. The reaction in the right-hand cell is

$$3\text{AgCl} + \text{La} + 3\text{F}^- \rightarrow \text{LaF}_3 + 3\text{Ag} + 3\text{Cl}^-$$

Hence for such a concentration cell without transport

$$E = \frac{RT}{F} \ln \frac{a_{\text{F}^-,2} a_{\text{Cl}^-,1}}{a_{\text{F}^-,1} a_{\text{Cl}^-,2}}$$

If the activity (concentration) of chloride ions on both sides of the lanthanum fluoride crystal is the same

$$E = \frac{RT}{F} \ln \frac{a_{\text{F}^-,2}}{a_{\text{F}^-,1}} \tag{85}$$

and E is positive if $m_2 > m_1$. Since m_1, the concentration of the reference solution, is constant Eq. (85) becomes

$$E = \text{constant} + \frac{RT}{F} \ln a_{\text{F}^-}$$

Similar equations can be obtained for all ion-selective electrodes. The properties of the ion exchanger do not appear in the net cell reaction and do not need to be known in detail for thermodynamic purposes.

No electrode is entirely selective towards the specified ion, and the presence of other ions can seriously impair the electrode performance. An empirical equation of the form

$$E = \text{constant} \pm \frac{RT}{F} \ln (a_i + K_{i,j} a_j)$$

has been proposed, where i and j are two singly charged ions and $K_{i,j}$ is the selectivity coefficient, the positive sign is taken if i and j are cations and negative if anions. For response to ion i only, $K_{i,j}$ must be small. The selectivity coefficient can be determined by maintaining the concentration of interfering ion constant in one solution and zero in the other. The dissolution of active material, poisoning, etc. can also interfere with the functioning of these electrodes.

Ion-selective electrodes find many applications: for example, in titrations (EDTA, halide, mixed halide, sulphate), water hardness determinations, analysis and control of plating baths, gas determinations. Care must be taken in the selection and use of such electrodes to avoid possible interference by other ions.

Further reading
A. K. Covington (ed.), *Ion-Selective Electrode Methodology*, vol. 1, CRC Press, Cleveland (1979)
H. Frieser (ed.), *Ion-Selective Electrodes in Analytical Chemistry*, vol. 1, Plenum Press, New York (1978)
H. Frieser (ed.), *Ion-Selective Electrodes in Analytical Chemistry*, vol. 2, Plenum Press, New York (1980)
N. Lakshminarayanaiah, Membrane phenomena, in *Electrochemistry*, vol. 4, Specialist Periodical Reports, Chemical Society, London, p. 167 (1974)
M. E. Meyerhoff and Y. M. Franticelli, *Anal. Chem.*, **54**, 27R (1982)
R. L. Solsky, Ion selective electrodes in biomedical analysis. *CRC Crit. Rev. Anal. Chem.*, **14**, 1 (1983)

Ion-selective field effect transistor (ISFET). A hybrid of an **ion-selective electrode** (ISE) and a metal–oxide field effect transistor (MOSFET). In a conventional potentiometric measurement, the signal from the ISE is trans-

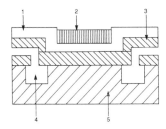

Figure 81. Cross-section of an ISFET: 1, silicon nitride layer; 2, PVC membrane impregnated with ion-selective component; 3, silicon oxide; 4, *n*-doped zone; 5, silicon substrate

174 Ion-selective membrane

mitted by wire to the input MOSFET of the voltmeter, thereby modulating the drain current. In an ISFET, the metal gate of the MOSFET is replaced by, or contacted with, a solid or liquid ion-sensitive membrane (see Figure 81). This development opens several possibilities, such as miniaturization, all solid-state design and *in situ* signal processing.

The response of such sensors is linked to a current rather than to a potential as in a conventional ion-selective electrode. Although a reference electrode may not be needed because of the presence of impurities in the gate, the use of a reference electrode is recommended because it results in a more stable electrical signal.

There are three distinct periods in the operational life of an ISFET: (1) pre-conditioning period (0–100 h); (2) working period (100–300 h); (3) terminal period (>300 h). During the first and last periods, the device is less stable as revealed by drifts in the response. The Nernstian response and selectivity decreases when the device is stored in solution; it is best stored above, rather than in, a solution.

PVC membranes impregnated with silver halides, valinomycin, etc. have been tested. An obvious extension of their use is for the estimation of biological compounds by the incorporation of an enzyme preparation in the membrane.

Ion-selective membrane. Separator for cells which transports only anions or cations (or certain anions or cations). Solvent molecules may also be transported. Membranes may be inorganic superionic conductors (see **Solid electrolytes**) such as sodium β-alumina, or organic polyfluorinated polymers with side chains including acid groups. Two typical membranes of the latter type are Nafion, produced by E. I. Du Pont de Nemours & Co.,

$$(CF_2-CF_2-CF-CF_2)_n$$
$$|$$
$$(OCF_2-CF)_y-OCF_2CF_2SO_3H$$
$$|$$
$$CF_3$$

and Flemion (Ashai Glass Co.)

$$(CF_2-CF_2)_x-(CF_2-CF)_y$$
$$|$$
$$(OCF_2-CF)_m-O-(CF_2)_n-COOH$$
$$|$$
$$CF_3$$

These membranes are in the form of thin sheets reinforced by plastic net.

Ion-selective membranes allow very thin cells to be constructed and are superior to asbestos and other physical separators. However, the resistance of

Ion-selective microelectrode

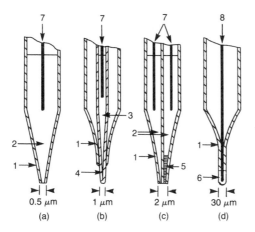

Figure 82. Cross-section through several types of microelectrodes: (a) reference electrode; (b) pH- or cation-selective glass electrode; (c) combined reference/liquid-exchanger electrode—now available in the form of four-barrelled microelectrodes for the simultaneous determination of potassium, sodium and calcium ions. 1, Glass capillary; 2, internal reference electrolyte; 3, internal electrolyte solution; 4, cation-sensitive glass; 5, liquid ion exchanger; 6, silver chloride deposited on silver-plated platinum wire; 7, silver–silver chloride electrode; 8, platinum wire

the membranes is relatively large allowing only low current densities in cells in which these operate. The inorganic separators are usually associated with high-temperature cells.

Ion-selective microelectrode (ISM). For intracellular measurements, an ISM has to penetrate the cell without causing undue damage; thus the effective tip should be as small as possible. When such an electrode is inserted into a cell, it records not only the potential difference which is created mainly by the concentration of the measured ion but it also records the cell membrane potential. When the reference electrode is as close as possible to the ISM, distortion of the electrical potential of the ISM by additional electrical activity (e.g., action or synaptic potentials) may be avoided. Double-barrelled glass micropipettes fulfil this requirement.

Several types of microelectrodes are available (*see* Figure 82); the liquid-exchanger microelectrodes have a higher selectivity compared to ion-sensitive glass electrodes.

Potassium-sensitive microelectrodes have been used in studies of the central nervous system to provide detailed information about dynamic changes in ionic composition of extracellular and intracellular spaces under various physiological and pathophysiological conditions. Potassium-, sodium-, calcium- and chloride-selective microelectrodes have been used for extracellular measurements in the brain. The potassium-sensitive electrode has been most extensively used for measuring potassium ion transients in the spinal cord; potassium ions accumulate in the extracellular space during stimulation of the sciatic nerve.

ISM with liquid ion exchangers have been used to measure ionic transients in various sensory organs and in molluscan single neurones. See Koryta (1980), chapter 5.

Further reading
H. J. Berman and N. C. Herbert (ed.), 'Ion-selective micro electrodes' in *Advances in Experimental Medicine and Biology*, vol. 50, Plenum Press, New York (1974)

ISFET. *See* **Ion-selective field effect transistor.**

ISM. *See* **Ion-selective microelectrode.**

Isoelectric focusing. The migration of ampholytes (e.g., proteins) in a pH gradient under an applied potential gradient. Molecules possessing a net positive or negative charge migrate in the electric field towards the region in which they are isoelectric. At their **isoelectric point**, the molecules have no charge and do not migrate in an electric field; they therefore concentrate into a narrow zone. A steady-state equilibrium is attained between migration in the electric field towards the isoelectric point and diffusion in the opposite direction. Separation of components is solely in respect of only one parameter: the isoelectric point.

If an electric field is applied to a mixture of proteins A and B at pH 6 in a linear pH gradient (*see* Figure 83), protein A ($pI = 5$) will be negatively charged and move towards the positive electrode at an ever-decreasing velocity until it reaches its isoelectric point where it focuses. Similarly protein B, initially at a pH below its isoelectric point ($pI = 8$) will be positively charged and will migrate towards the negative electrode until it focuses in the region of pH 8. In general, the point of application of the sample is immaterial, each protein will focus at its own pI. One of the great advantages of electrofocusing is the constant counteraction of diffusion. The resolution is easily varied by the pH range used; proteins with as small a difference of 0.02 pH units in their isoelectric point can be separated.

The success of the separation depends on the establishment and maintenance of a near linear pH gradient and a uniform conductance along the gradient during an experiment. The components used to form the gradient are of major importance. The carrier ampholytes must have a good buffering capacity and solubility in water at their isoelectric point, with low relative molecular mass in comparison to proteins and a low absorption of ultraviolet radiation at 280 nm. Synthetic ampholytes, with a range of pI values, are the

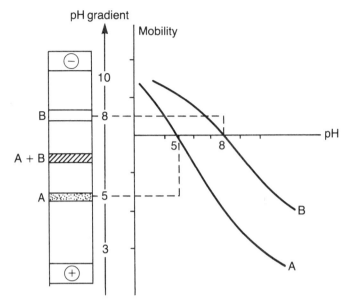

Figure 83. Principles of isoelectric separation of two proteins A and B

most useful substances available commercially: for example, the Ampholines (LKB-Produkter) polyamino-polycarboxylic acids, Servalyt (Serva) a similar material containing sulphonic acid and phosphoric acid groups and Pharmalyte (Pharmacia) copolymers of epichlorohydrin with amines, amino acids and peptides.

Electrofocusing columns are usually stabilized with a sucrose density gradient (in the range 5–50 percent w/v). The electrolyte around the lower positive electrode is acidic (e.g., sulphuric or phosphoric acid) and is 60 percent w/v sucrose; and around the upper negative electrode the electrolyte is alkaline (e.g., sodium hydroxide or ethanolamine). Under certain circumstances, especially with narrow range columns, it is advisable to reverse the electrodes to ensure uniformity of conductivity.

Electrofocusing in gel media, as compared with density gradients, imposes much greater demands on the uniformity of electrical conductance; in density columns, localized heating is overcome by mixing, but this is not possible in gels. Stabilizing media must be capable of preventing convection, supporting the focusing zones and of maintaining a stable pH gradient. Electroosmosis must be kept to a minimum otherwise the resulting flow will cause the movement of the whole solution. Polyacrylamide and agarose gels, chemically treated cellulose acetate and Sephadex containing carrier ampholytes have been used as supporting media. The apparatus used for electrofocusing in stabilizing media is similar to that used for zone electrophoresis (*see* Figure

66). The width (x) of the Gaussian distribution of a focused zone is given by the theoretical equation

$$x = \left[\frac{D}{X(du/dpH)(dpH/dx)}\right]^{1/2}$$

where D is the diffusion coefficient, X the applied field strength, du/dpH the rate of change of mobility with pH (i.e. the slope of the mobility–pH plot) and dpH/dx the pH gradient. A modest improvement in the resolution can be achieved by decreasing the pH gradient by, for example, using a narrower range of carrier ampholytes or by varying the thickness of the gel. Heat dissipation is the limiting factor in attempting to increase the resolution by increasing the field strength. More efficient heat removal is achieved with ultra-thin gel films, which have the additional advantages that they allow more rapid fixing and staining of the protein zones.

Although this separation method is capable of handling quite large sample loads, smaller loads lead to higher resolution. Once the separation has been achieved, the field is switched off and the zones revealed by such methods as staining with dyes, silver salts or by immunofixation.

Two-dimensional separation, on a square plate, in which isoelectric focusing is achieved in one direction and subsequently a second separation by gel electrophoresis in the direction at right angles, gives rise to a much improved separation. Although 1100 different proteins have been shown to be present in extracts from *Escherichia coli* and over 300 from plasma, there are problems with the actual identification of the individual spots.

Preparative procedures for isoelectric focusing are now available commercially. There is no restriction to the scaling up. The temperature, pH, conductivity and absorption at 280 nm are continuously monitored and separate fractions collected.

Further reading
P. G. Righetti, E. Gianazza and B. Bjellqvist, Modern aspects of isoelectric focusing, *J. Biochem. Biophys. Methods*, **8**, 89, 109, 135 and 157 (1983)
C. F. Simpson and M. Whittaker (ed.), *Electrophoretic Techniques*, Academic Press, London (1983)

Isoelectric point (pI). For an amino acid or protein, the isoelectric point is the pH at which the species does not move in an electric field. Since the isoelectric point varies with the environment, it is necessary to specify the nature of the **buffer solution** and its **ionic strength**. The isoelectric point may be determined experimentally by measuring the electrophoretic mobility (*see* **Electrophoresis**) in a buffer solution of fixed ionic strength over a range of pH values and observing the pH at which charge reversal (i.e. zero net charge) occurs.

For amino acids, the p*I* may also be calculated from the dissociation constants of the amino and carboxyl groups as follows:

For monoamino-monocarboxylic acids $pK = \frac{1}{2}(pK_1 + pK_2)$
For monoamino-dicarboxylic acids $pK = \frac{1}{2}(pK_1 + pK_2)$
For diamino-monocarboxylic acids $pK = \frac{1}{2}(pK_2 + pK_3)$

Reducing the net charge on a protein molecule results in the collapse of the electric double layer and decreases the stability of the protein. At the isoelectric point, a protein is most sensitive to precipitation by species with affinity for the water sheath that surrounds the molecule (e.g., ethanol, acetone and certain neutral salts). Typical values for the isoelectric points of amino acids and proteins are tabulated in Table 23.

Table 23
Typical isoelectric points of amino acids and proteins

Amino acids	p*I*	Proteins	p*I*[a]
Aspartic acid	2.98	Bovine serum albumin	4.9
Glutamic acid	3.22	β-Lactoglobulin	5.2
Serine	5.68	Carboxypeptidase	6.0
Glycine	5.97	Haemoglobin	6.7
Alanine	6.02	Sickle cell haemoglobin	6.9
Histidine	7.59	Ribonuclease	9.5
Lysine	9.74	Cytochrome *c*	10.7
Arginine	10.76	Lysozyme	10.7

[a] The precise value depends on the temperature and the ionic strength of the medium.

Isopotential point. The point at which a surface (e.g., a colloidal particle) is completely discharged, the **electrical double layer** has collapsed, and the ζ-**potential** and surface charge are zero. With increasing concentration of electrolyte, the ζ-potential at an interface decreases approaching zero. The concentration of electrolyte at which the isopotential point is attained depends on the valence type of the electrolyte. *See also* **Potential of zero charge.**

Isotachophoresis. An electrophoretic separation method in which a steady-state configuration is obtained according to the moving boundary principle (*see* **Transport number**).

Consider the separation of the anionic species A, B and C in a narrow bore

180 Isotachophoresis

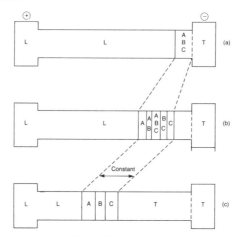

Figure 84. Isotachophoretic principle for separation of a mixture of anions $(u_L > u_A > u_B > u_C > u_T)$

tube (see Figure 84(a)) in which the tube and anode compartment are filled with the 'leading' electrolyte (L) and the cathode compartment is filled with the 'terminating' electrolyte (T). The mobility of the anions of L (T) must be greater (less) than the mobilities of A, B and C (i.e. $u_L > u_A > u_B > u_C > u_T$).

When a current is passed through such a system each ion will move at a different velocity ($v_i = u_i X$); the ion with the highest mobility will run forward and those with lower mobilities will lag behind. Since the anions of the leading electrolyte can never be overtaken, neither can the anion of the terminating electrolyte overtake A, B or C, the anions A, B and C are sandwiched between L and T. In the mixed zone (see Figure 84(b)), the separation continues and after some time a series of zones is obtained (in order of decreasing mobility) in which each zone contains only one component (see Figure 84(c)) again sandwiched between L and T. When this steady state has been achieved, all the zones run together at equal velocities, thus

$$v_L = v_A = v_B = v_C = v_T$$

or

$$u_L X_L = u_A X_A = u_B X_B = u_C X_C = u_T X_T$$

Since the species are arranged in order of increasing mobility, it follows that, working at constant current density, the field strength increases and there is an increase in the heat generated and hence an increase in temperature (see Figure 85). When the steady state has been attained, the zone boundaries are self-correcting and will not broaden (compare with zone electrophoresis where the zones are broad owing to diffusion and absorption). If an ion lags

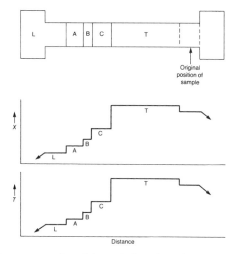

Figure 85. Schematic representation of the variation of applied field (X), and temperature (T) for different zones moving in steady state

behind in a zone with a higher field strength, it will acquire a higher velocity and move forward into its correct zone.

The concentrations of the species in the individual zones as they migrate is fixed by the concentration in the preceding zone according to the Kohlrausch regulating function

$$\frac{u_L}{u_A} = \frac{c_L}{c_A}, \quad \frac{u_A}{u_B} = \frac{c_A}{c_B}, \quad \text{etc.}$$

Thus for a very dilute sample, not only is there a separation into the various components but there can be an appreciable concentration depending on the concentrations of the leading and terminating electrolytes.

Methods available for the detection of the zones in the separations include fixed point thermocouples (to monitor the change of temperature as the zones pass), electrodes (to monitor changes in the conductance or mobility of the ions in each zone) and for species that absorb in the ultraviolet region an ultraviolet detector. Analytical isotachophoresis has been applied to the analysis of inorganic anions and cations, the purity of drugs and the identification of uric acid, purine and pyrimidines in serum.

Capillary isotachophoresis can be used as a preparative method; two main modifications to the analytical equipment are necessary (*see* Figure 86). A membrane (M_2) is included after the terminating electrolyte reservoir. Immediately after the detector (D), two T-pieces are inserted to permit the separated zones to be swept out by the leading electrolyte. Since the sample zones may be very small (about 10–20 pm^3), care has to be taken to ensure that

182 Isotachophoresis

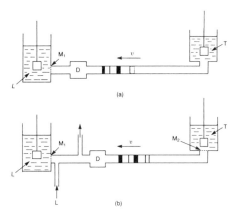

Figure 86. Schematic representation of the apparatus for (a) analytical capillary isotachophoresis and (b) preparative capillary isotachophoresis

the velocity of the eluting stream is only about 10 percent greater than migrating velocity otherwise the zones will be diluted. The zones can be collected on a cellulose acetate strip.

Further reading

F. M. Evaraerts, J. L. Beckers and T. P. E. M. Verheggen, Isotachophoresis, theory and instrumentation, *Journal of Chromatography Library*, vol. 6, Elsevier, Amsterdam (1976)

C. F. Simpson and M. Whittaker (ed.), *Electrophoretic Techniques*, Academic Press, London (1983).

K

Kohlrausch's law. *See* **Conductivity at infinite dilution; Molar ionic conductivity.**

L

Lead–acid battery. Introduced by Plante in 1860 as an electrochemical storage device. The reactions occurring at each electrode are

cathode $\quad PbO_2 + 3H^+ + HSO_4^- + 2e \underset{\text{charge}}{\overset{\text{discharge}}{\rightleftharpoons}} PbSO_4 + 2H_2O \quad$ (a)

anode $\quad\quad\quad\quad\quad\quad\quad Pb + HSO_4^- \underset{\text{charge}}{\overset{\text{discharge}}{\rightleftharpoons}} PbSO_4 + H^+ + 2e \quad$ (b)

overall $\quad Pb + PbO_2 + 2H^+ + 2HSO_4^- \rightleftharpoons 2PbSO_4 + 2H_2O \quad$ (c)

The electrodes are usually made of a grid of a lead–antimony alloy (which gives mechanical strength), filled with a paste of litharge and red lead in sulphuric acid. Fillers may be added to mitigate the disruptive effects of the large volume changes that accompany the cell reaction, and soluble substances to increase the porosity. When these plates are electrolysed in sulphuric acid, the charging process (reaction (a)) converts the anode material into porous lead dioxide, and the cathode material is reduced to spongy lead (reaction (b)). Each cell of a lead storage battery usually contains a cathode consisting of a number of such plates connected in parallel and, slotting into these, a similar arrangement of anode plates, each pair of plates being separated by a thin porous sheet of glass fibre or some plastic material.

The electrolyte is an aqueous solution of sulphuric acid. This is consumed in the discharge reaction, and enough must be present to provide for this and also for a sufficient excess to give good conductivity in the discharged cell. The more concentrated the acid used, the lighter will be the cell; but at very high concentrations, the conductivity decreases and the freezing point rises. The optimum value is about 35 percent sulphuric acid by weight, and the conductivity is then near its maximum throughout the charge–discharge cycle. The specific gravity is about 1.26 falling to about 1.1 when the cell is discharged. This gives a simple way of checking the state of charge of the cell.

The equation for the cathodic reaction can be written

$$PbO_2 + 4H^+ \rightleftharpoons Pb^{4+} + 2H_2O$$
$$Pb^{4+} + 2e \rightleftharpoons Pb^{2+}$$
$$Pb^{2+} + HSO_4^- \rightleftharpoons PbSO_4 + H^+$$

The potential difference corresponding to the electrochemical step is

$$E = E^{\ominus}_{Pb4+,Pb2+} + \frac{RT}{2F} \ln \frac{a_{Pb4+}}{a_{Pb2+}}$$

The activities of lead (IV) and lead (II) ions are very small and nearly equal, so the reversible potential +1.74 V, is practically that of the standard lead (IV)–lead (II) electrode.

At the other electrode, the reaction $Pb \rightleftharpoons Pb^{2+} + 2e$ essentially determines the electrode potential which is

$$E = E^{\ominus}_{Pb2+,Pb} + \frac{RT}{2F} \ln a_{Pb2+}$$

The concentration of lead (II) ions in sulphuric acid saturated with lead (II) sulphate is 5×10^{-6} mol dm^{-3}, which corresponds to a value of -0.28 V for the anode potential. The theoretical voltage is therefore $1.74 + 0.28 = 2.02$ V, which is very close to the values found.

Typical curves for the charge and discharge of a lead accumulator are shown in Figure 87. Charging proceeds at an almost steady voltage after an initial rise, but eventually the voltage turns rapidly upwards and reaches the value at which hydrogen and oxygen are evolved freely. This marks the completion of charging, and continued evolution of gas is not only wasteful of energy but may assist the disintegration of the plates. During discharge, the e.m.f. falls quickly to just below 2 V, and then gradually to 1.8 V. The cell should be recharged before the final drop in voltage is reached.

The theoretical **energy density** is 161 W h kg^{-1} and power density 84 A h kg^{-1}. In practice, the values of these parameters are much lower and compare unfavourably with the values for other battery systems: for example, **nickel–cadmium battery, sodium–sulphur battery** and **metal–air battery**. The current

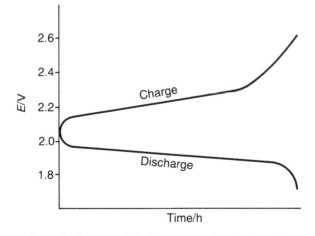

Figure 87. Charge and discharge curves for a lead–acid battery

efficiency of a lead–acid battery (i.e. the ratio of the quantity of electricity obtained from the battery to the quantity of electricity required for charging) is higher than 90 percent. The energy efficiency (*see* **Electrochemical storage**) is lower (75–80 percent) because of the interval between the charging voltage and discharging voltage.

The internal resistance of the cell is very small, and the main reason for the loss in efficiency is concentration **overpotential**, especially when the pores of the electrode are partially blocked by the lead sulphate precipitate.

A charged accumulator does not retain its charge indefinitely. The loss of charge can be put roughly at an average of 1 percent per day. The loss can be attributed to a variety of causes. At the positive plate, lead dioxide can be lost by local action with the lead of the grid; electrochemical corrosion will also occur at the site of any metal impurity in the lead electrode. To minimize this danger, distilled water must be used to top up the electrolyte.

When a lead–acid battery is operating, lead sulphate is deposited in the pores and at the surfaces of the electrodes in a very finely divided form. On standing, this will tend to consolidate into larger crystals which can block the pores, thus reducing the electrode surface, or can be shed from the electrodes and fall to the bottom of the cell. This is known as sulphating and is the chief cause of the deterioration of cells. To minimize this effect, a cell should not be allowed to stand in a partially discharged state.

Despite the problems discussed above, the lead–acid battery has maintained its popularity where high discharge currents, long life and overall robustness is called for.

Further reading
K. V. Kordesch (ed.), *Batteries*, vol. 2, *Lead Acid Batteries and Electric Vehicles*, Marcel Dekker, New York (1977)

Leclanché cell. A primary dry cell developed in the 1880s by Leclanché. It consists of an ammonium chloride solution into which dips a zinc rod as the anode. The cathode, in a porous container, consists of a carbon rod packed round with powdered manganese dioxide and carbon. The anode reaction is

$$Zn \rightarrow Zn^{2+} + 2e$$

and at the cathode

$$MnO_2 + H_2O + e \rightarrow MnO(OH) + OH^-$$

The zinc ions react with the alkaline ammonium chloride and a zinc ammine crystallizes out

$$Zn^{2+} + 2OH^- + 2NH_4^+ \rightarrow Zn(NH_3)_2^{2+} + 2H_2O$$

so that the overall reaction is

$$Zn + 2MnO_2 + 2NH_4Cl \rightarrow Zn(NH_3)_2Cl_2 + 2MnO(OH)$$

The cathode reaction may consist of a primary electrochemical step

$$MnO_2 + 4H^+ + 2e \rightarrow Mn^{2+} + 2H_2O$$

followed by chemical steps resulting in the reaction

$$Mn^{2+} + MnO_2 + 2OH^- \rightarrow 2MnO(OH)$$

In the common dry cell, the electrolyte is a concentrated ammonium chloride solution containing some zinc chloride and a small amount of mercury (II) chloride. The latter is reduced to mercury at the zinc surface which reduces the corrosion rate and adds to the shelf life of the cell. The usual form of the cell is a zinc container, acting as anode next to which is a layer of the electrolyte mixture made into a gelled paste by the addition of starch. Inside this is the manganese dioxide/carbon mixture surrounding a central carbon rod; the latter is fitted with a metal cap to act as positive terminal, and the cell is sealed with a bitumen layer. The Leclanché cell may be constructed with a capacity of 1–10 A h, although at high current drain only a quarter of the rated capacity may be used. It has a good **energy density** (55–$77\,W\,h\,kg^{-1}$, 120–$152\,W\,h\,dm^{-3}$) and is particularly cheap to make. Its discharge performance is inferior to that of an **alkaline cell** (*see* Figure 6).

A similarly constructed cell uses magnesium in place of zinc, with a magnesium bromide electrolyte. Magnesium has a reversible potential 1.5 V more negative than that of zinc, but it is a mixed potential that controls the electrode in aqueous medium, and the cell gives about 1.9 V compared with about 1.6 V for the Leclanché cell.

Levich equation. *See* **Rotating-disc electrode.**

Limiting current density. *See* **Diffusion-limited current.**

Linear sweep voltammetry. A technique in which the voltage is scanned at a constant rate between two voltage limits while the current is displayed on a recorder or oscilloscope. If the voltage is scanned back and forth between the limits, the technique is known as cyclic linear sweep voltammetry or simply cyclic voltammetry. A complete theoretical analysis of the method is complex. However, information about fast adsorption kinetics may be obtained and cyclic voltammetry is often used for a preliminary investigation of surfaces.

The current passed as the voltage is swept is made up of faradaic and capacitative components. For sweep rates less than $1\,V\,s^{-1}$, the capacitative current is usually small compared to the faradaic current, which is for a reaction producing an adsorbed intermediate

188 Linear sweep voltammetry

$$I = AC_D\left(1 - \frac{\int I dt}{q_m}\right) \exp\left[\frac{\beta F}{RT} V(t)\right]$$

C_D is the **pseudocapacitance** of the adsorbed layer and q_m is the charge required to form a monolayer. As the coverage of the intermediate (e.g., a surface oxide) goes from zero to one, the current passes through a maximum giving a characteristic peak. If the adsorption is reversible, the potential of maximum current of the oxidation and reduction peaks coincides and is independent of scan speed. Irreversible behaviour is manifested as asymmetry in the oxidation and reduction peaks, which becomes more pronounced with increasing sweep rate.

Figure 88 is a typical cyclic voltammogram of the current at a metal electrode as it is swept from just negative of the hydrogen-evolution potential to a potential at which oxygen is evolved. Peaks due to the adsorption and oxidation of hydrogen atoms are seen near 0 V (versus RHE). The electrode also shows surface oxidation prior to oxygen evolution (*see* **Oxygen electrode reactions**) which is reduced as the potential is swept to more negative voltages.

Cyclic voltammetry has been used to characterize the surfaces of heterogeneous catalysts, giving similar information to temperature-programmed desorption or temperature-programmed oxidation or reduction.

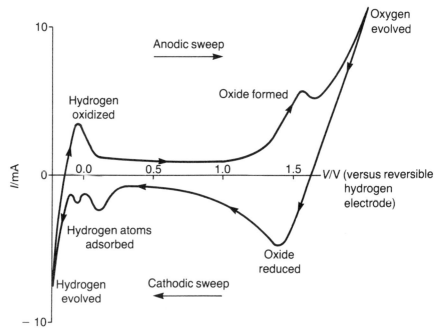

Figure 88. Cyclic voltammogram of platinum in alkaline solution at a sweep rate of 40 mV s^{-1}

Liquid junction potential. The potential difference established across the interface between two dissimilar electrolyte solutions. The potential, due to the diffusion of ions across the interface, acts to retard the more rapidly diffusing ions and to accelerate the more slowly diffusing ions, whether they are cations or anions. In this way, equilibrium is soon established and a steady liquid junction potential is set up; the magnitude (<0.1 V) depends on the **transport number** of the ions, the charge they carry and the concentration of the electrolyte.

The e.m.f. of the **concentration cell**

$$\text{Pt}, \text{H}_2(g) \quad | \quad \text{HCl} \quad \vdots \quad \text{HCl} \quad | \quad \text{H}_2(g), \text{Pt}$$
$$(101\,325 \text{ N m}^{-2}) \quad (a_2) \quad \vdots \quad (a_1) \quad (101\,325 \text{ N m}^{-2})$$

given by

$$E_{\text{cell}} = \frac{2t_- RT}{F} \ln \frac{a_{\pm,1}}{a_{\pm,2}}$$

is the sum of two electrode potentials and the liquid junction potential, E_j. The algebraic sum of the two electrode potentials is, theoretically,

$$E_1 - E_2 = \frac{RT}{F} \ln \frac{a_{H+,1}}{a_{H+,2}}$$

Hence,

$$E_j = E_{\text{cell}} - (E_1 - E_2)$$

$$= \frac{2t_- RT}{F} \ln \frac{a_{\pm,1}}{a_{\pm,2}} - \frac{RT}{F} \ln \frac{a_{H+,1}}{a_{H+,2}}$$

Assuming the ratio $a_{H+,1}/a_{H+,2} = a_{\pm,1}/a_{\pm,2}$, then

$$E_j = \frac{(2t_- - 1)RT}{F} \ln \frac{a_{\pm,1}}{a_{\pm,2}} = \frac{(t_- - t_+)RT}{F} \ln \frac{a_{\pm,1}}{a_{\pm,2}}$$

or, in general, for cation-reversible electrodes

$$E_j = \left(t_- \frac{\nu}{\nu_+} - 1 \right) \frac{RT}{nF} \ln \frac{a_{\pm,1}}{a_{\pm,2}}$$

and for anion-reversible electrodes

$$E_j = \left(t_+ \frac{\nu}{\nu_-} - 1 \right) \frac{RT}{nF} \ln \frac{a_{\pm,1}}{a_{\pm,2}}$$

For cation-reversible electrodes when $t_+ = t_-$, $E_j = 0$; when $t_- > t_+$ E_j is positive and adds to the sum of the electrode potentials; and when $t_+ > t_-$ E_j is negative and yields a lower e.m.f. for the cell than the sum of the electrode potentials. Attempts have been made to eliminate liquid junction potentials by interposing a salt bridge consisting of a concentrated solution of potassium chloride, potassium nitrate, etc., for which $t_+ \approx t_-$. As the solution is so concentrated, most of the diffusion is due to the bridge electrolyte. Thus the junction potential of the original cell is replaced by two junction potentials acting in opposition with values very near zero. Junction potentials may be reduced in this way, but it is doubtful if they are completely eliminated.

Liquid membrane. *See* **Ion-selective electrode.**

Lithium electrometallurgy. Lithium is produced by an electrolytic process broadly similar to that used for sodium (*see* **Sodium electrometallurgy**). The electrolyte is a fused mixture of lithium and potassium chlorides at about 680 K. Lithium is preferentially liberated at the cathode; its standard potential in water ($E^\ominus_{Li^+,Li} = -3.05$ V) is more negative than that of potassium ($E^\ominus_{K^+,K} = -2.93$ V), but in the absence of water the order is reversed. Morever, the conductivity of lithium chloride is nearly four times that of potassium chloride, and the complexing of the latter in a mixture would result in a much higher activity of the lithium ion. The electrolysis is conducted at about 8 V, and the current efficiency is 85–90 percent.

Luggin capillary. When making electrochemical measurements using a **three-electrode system**, the potential measured between the working and reference electrodes contains a component due to the potential drop developed when the polarizing current overcomes the resistance of the electrolyte. This so-called *IR* drop may be minimized by constricting the flow of ions between the working and reference electrodes by means of a thin capillary. The only contribution to the *IR* drop now comes from the solution between the capillary (the Luggin capillary) and the working electrode. This distance is made as small as possible, usually down to the diameter of the capillary (*see* Figure 89). If the Luggin capillary is brought too close to the electrode surface, its presence disrupts the distribution of current at the electrode. The *IR* drop is measured by the **interruptor technique**.

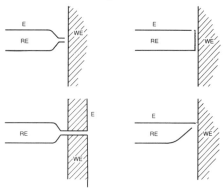

Figure 89. Different types of Luggin capillary used in measurements of potential. RE, reference electrode compartment; WE, working electrode; E, electrolyte

M

Magnesium electrometallurgy. Magnesium is obtained by electrolysing a fused mixture of magnesium, sodium and calcium chlorides in the approximate proportions 20:60:20. This mixture has a density greater than that of molten magnesium, and the cell is operated at a temperature somewhat above the melting point of magnesium (924 K), so that the molten metal floats and is removed from the top of the cell. The steel container acts as cathode, and a central graphite anode is contained in a porous pot. The chlorine that is liberated rises and is piped away for reuse in making the chloride.

The process is straightforward if pure anhydrous materials are used; the current efficiency is then over 90 percent and the product very pure. The main complication in the process, on account of the deliquescence of the starting materials, is contamination by water. Not only will any water present be electrolysed first, causing a drop in current efficiency, but it will also create magnesium oxide which sinks to the bottom taking some magnesium with it, thus causing an additional loss.

Membrane electrode. *See* **Biosensor; Gas-sensing membrane probe; Ion-selective electrode.**

Membrane potential. *See* **Donnan membrane equilibrium.**

Mercury cell. The overall reaction of the mercury dry cell is

$$M + HgO \rightarrow MO + Hg$$

where M is zinc or cadmium. The electrolyte is concentrated potassium hydroxide solution, and the metal ions dissolving from the anode are precipitated as the oxide and hydroxide. The cathode is a compressed mixture of mercury (II) oxide and graphite, and the cathode reaction is

$$HgO + H_2O + 2e \rightarrow Hg + 2OH^-$$

The electrolyte is not consumed in the reaction, so that little is necessary. The **energy density** of the cell is therefore high, and the cell can be made in very small sizes for calculators, etc. The cell has other advantages: the voltage (1.35 V for zinc, 0.9 V for cadmium) is stable, and it can give high currents without loss in performance; its shelf life is also high.

Mercury–mercury (I) salt electrode. Mercury (I) fluoride, bromide and iodide are well-characterized solids (isomorphous with calomel) which can be used as the basis of electrodes reversible to the corresponding halide species. The use of the fluoride electrode is restricted by the fact that mercury (I) fluoride is rapidly and completely hydrolysed in aqueous solution. It has been successfully used in liquid hydrogen fluoride in such cells as

$$\text{M} \mid \text{MF}_2 \text{ HF NaF} \mid \text{Hg}_2\text{F}_2, \text{Hg}$$

where M is copper, cadmium, silver or zinc. The bromide and iodide electrodes are photosensitive and all operations must be carried out in red light.

Extensive use has been made of mercury (I) acetate, oxalate, iodate, sulphate and oxide to produce electrodes reversible to the corresponding anion. See Ives and Janz (1961).

Standard electrode potentials are presented in Table 24.

Table 24
Standard electrode potentials at 298 K for mercury–mercury (I) salt electrodes

Electrode	$E^{\ominus}$/V
$\text{OAc}^- \mid \text{Hg}_2(\text{OAc})_2, \text{Hg}$	+0.5113
$\text{Br}^- \mid \text{Hg}_2\text{Br}_2, \text{Hg}$	+0.1396
$\text{Cl}^- \mid \text{Hg}_2\text{Cl}_2, \text{Hg}$	+0.2682
$\text{I}^- \mid \text{Hg}_2\text{I}_2, \text{Hg}$	−0.0405
$\text{SO}_4^{2-} \mid \text{Hg}_2\text{SO}_4, \text{Hg}$	+0.6158
$\text{IO}_3^- \mid \text{Hg}_2(\text{IO}_3)_2, \text{Hg}$	+0.3944
$\text{Ox}^{2-} \mid \text{Hg}_2\text{Ox}, \text{Hg}$	+0.4166
$\text{OH}^- \mid \text{HgO}, \text{Hg}$	+0.9260

Mercury–mercury oxide electrode. *See* **Mercury–mercury (I) salt electrode; Oxide electrode.**

Metal–air battery. Primary or secondary battery in which the cathode reaction is the reduction of atmospheric oxygen

$$O_2 + 2H_2O + 4e \rightarrow 4OH^-$$

with the oxidation of the metal at the anode

$$M \rightarrow M^{z+} + ze$$

where M is zinc, magnesium or aluminium.

The battery may be recharged by plating the metal, with oxygen evolution at a third electrode, or simply be mechanically recharged by the addition of a replacement anode. The construction of the air cathode is usually of high surface area carbon with a hydrophobic binder. This allows oxygen to permeate the electrode, but does not allow the escape of electrolyte. The **energy density** achieved by zinc–air batteries is theoretically about 500 W h kg^{-1}. The limiting rate of discharge depends on the rate of oxygen reduction. For high discharge rates, air must be blown across the cathode. A zinc–air **fuel cell** may be constructed in which the 'fuel', a metallic zinc slurry, is cycled through the cell. The spent fuel may be reduced or replaced as in the case of the battery.

Metal electrode. An electrode reversible to the metal ions in solution

$$M \rightleftharpoons M^{z+} + ze$$

The **electrode potential** is given by

$$E_{M^{z+},M} = E^{\ominus}_{M^{z+},M} + \frac{RT}{zF} \ln a_{M^{z+}}$$

Metal–metal oxide electrode. *See* **Antimony–antimony oxide electrode; Mercury–mercury (I) salt electrode; Oxide electrode.**

Metal winning. *See* **Electrowinning.**

Micelles. *See* **Colloidal electrolytes.**

Microbial fuel cell. A fuel cell is a device for the direct conversion of chemical energy into electrical energy. It requires two electrodes, a supporting electrolyte solution and an external circuit to use the electricity. Inorganic fuel cells use hydrogen as the anodic reactant and oxygen as the cathodic reactant.

In microbial fuel cells, microorganisms have been used in the anode compartment in two ways. In the 'direct' microbial fuel cell, oxidative degradation reactions within the living cell provide the reducing power to generate an anodic potential, whereas in the 'indirect' cell the microorganisms generate a fuel (e.g., hydrogen) which is then utilized as a reductant in the electrochemical reaction at the anode. The performance of direct cells is poor, because the cell wall screens the respiratory processes of the cell from the electrodes. The presence of a redox dye increases the rate of electron transfer from the microorganisms to the anode. This in turn results in an improvement in the performance of such cells and may make them practical as low-power sources of electricity.

To obtain maximum power and optimum yields of electricity, consideration has to be given to: (1) selection of the fuel (carbon source) and organism; (2) choice of cathode; (3) identification of redox mediators which are rapidly reduced by the organisms; (4) rapid and reversible oxidation of the mediator at the anode; and (5) the cell design to reduce overpotential and internal resistance.

The choice of organisms depends largely on the substrate available for oxidation; the following substrates and organisms have been used: glucose with *Escherichia coli, Pseudomonas aeruginosa* and *Proteus vulgaris*, succinate with *Alcanigenes eutrophus*, sucrose with *Saccharomyces cerevisiae* and lactose with *E. coli* ML308. The cathode is usually platinum in a **catholyte** solution which contains potassium ferricyanide; this provides a stable relatively non-polarizable cathodic reaction. Potassium ferricyanide is a good electron acceptor when used as an alternative to oxygen under anaerobic conditions. A relatively low e.m.f. is produced from such cells because of the high standard electrode potential of the $Fe(CN)_6^{3-}|Fe(CN)_6^{4-}$ couple ($E^\ominus = 0.36$ V). Mediators with electrode potentials less than 0 V are required to give cells of appreciable voltage and useful power. Thionine has proved to be a good mediator; it is reduced by *E. coli* at rates comparable to the rate of electron transfer by oxygen; *P. vulgaris* reduces thionine at twice this rate.

In a typical microbial fuel cell (shown schematically below) operating under anaerobic conditions

Anode	Anolyte	Catholyte	Cathode
$\ominus$			$\oplus$
	Phosphate buffer soln	Phosphate buffer soln	
	Thionine	$K_3Fe(CN)_6$	
Platinized carbon	Carbon source Microorganism		Platinum
	(oxygen-free nitrogen bubbled through both compartments)		
	Cation-permeable ion-exchange membrane (Nafion)		

the electrode reactions may be written as

microorganism (reduced intermediates) + oxidized form of mediator $\rightleftharpoons$ microorganism (oxidized intermediates) + reduced form of mediator

reduced form of mediator $\rightleftharpoons$ oxidized form of mediator + $2H^+ + 2e$

$$Fe(CN)_6^{3-} + e \rightleftharpoons Fe(CN)_6^{4-}$$

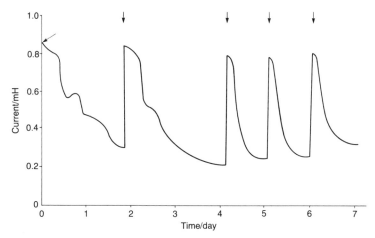

Figure 90. Demonstration of the regenerative capacity of microbial fuel cell. *E. coli* ML 318 (200 mg dry weight) metabolizing lactose. Arrows indicate the addition of lactose (10 μmol) and the restoring of the anolyte to pH 7

For such a cell (20 cm^3 volumes of electrolyte in each compartment) using succinate and 1 g dry weight of *A. eutrophus*, an open circuit voltage of 0.7 V has been obtained with a current of 2–3 mA at 0.5 V for several hours or 30 mA at 0.4 V for a few minutes. The rate of thionine reduction by *E. coli* metabolizing glucose is equivalent to a rate of energy transfer of about 1 W per g dry weight of cells; higher concentrations of the mediator may increase this. Figures 90 and 91 demonstrate the regenerative nature of microbial fuel cells.

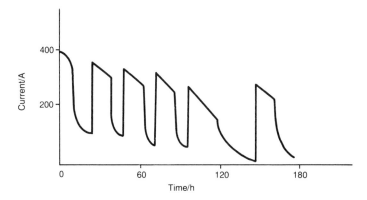

Figure 91. Demonstration of the regenerative capacity of a microbial fuel cell. Cells of *E. coli* suspended in phosphate buffer solution at pH 7.0 using thionine as mediator. Additions of 20 μmol of glucose were made and the cell allowed to run down under load over a period of 24 h. The cells were then refed with the original amount of glucose without removing the bacteria, and allowed to run down. On the sixth day the cell was starved and then fed again on day 7. (From Stirling *et al.* (1983))

196 Microelectrode

Similar cells in which specific enzymes are used instead of intact cells have been described.

Further reading

H. P. Bennetto, J. L. Stirling, K. Tanaka and C. A. Vega, Anodic reactions in microbial fuel cells, *Biotechnol. Bioengng*, **25**, 559 (1983)

J. L. Stirling, H. P. Bennetto, G. M. Delaney, J. R. Mason, S. D. Roller, K. Tanaka and C. F. Thurston, Microbial fuel cells, *Biochem. Soc. Trans.*, **11**, 451 (1983)

L. B. Wingard, C. H. Shaw and J. F. Castner, Bioelectrochemical fuel cells, *Enzyme Microb. Technol.*, **4**, 137 (1982)

A. T. Yahiro, S. M. Lee and D. O. Kimble, Bioelectrochemistry. 1. Enzyme-utilizing bio-fuel studies, *Biochim. biophys. Acta*, **88**, 375 (1964)

Microelectrode. *See* **Ion-selective microelectrode.**

Mini-ion-selective electrode. An electrode with an outside diameter of about 1 mm with medical applications. Commercially available mini-**glass electrodes** are available for hydrogen, sodium and potassium ions. The electrodes have been miniaturized to a point where they fit into a 0.75 mm outside diameter injection needle (Figure 92(a)).

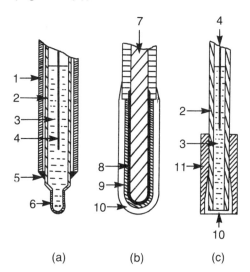

Figure 92. Cross-section of various types of mini-ion-selective electrodes: (a) pH electrode for catheter use; (b) coated wire electrode; (c) liquid membrane electrode. 1, steel capillary tubing (outside diameter 2 mm); 2, glass sheath; 3, internal filling solution; 4, internal reference electrode (silver–silver chloride); 5, epoxy bond; 6, pH glass membrane; 7, silver wire; 8, silver chloride layer deposited on silver electrode; 9, gelled contact electrolyte; 10, PVC membrane impregnated with ion-selective component; 11, polyethylene tubing

Liquid membrane (Figure 92(c)) electrodes (*see* **Ion-selective electrode**) offer more latitude in the preparation of miniaturized ion-selective electrodes. The coating of thin wires with solvent polymeric membrane material (Figure 92(b)) provides a ready miniaturization of sensors. The potential of such electrodes, however, varies with time because of the ill-defined half cell between the membrane adjacent to the metal and the metal itself.

Catheter tip electrodes, obtained by bonding valinomycin–PVC membranes into PVC tubing, permit the continuous monitoring of potassium ions in plasma and whole blood.

An attractive type of mini-electrode (currently under development) is the **ion-selective field effect transistor**. See Koryta (1980).

Mobility (u; units: m^2 s^{-1} V^{-1}). The velocity of an ion or colloidal particle under a unit potential gradient. It is related to the **molar ionic conductivity** (Λ_i) by the equation

$$\Lambda_i = u_i F$$

and in the limiting case by

$$\Lambda_i^\circ = u_i^\circ F$$

and to the **transport number** of an ion by

$$t_+ = \frac{u_+}{u_+ + u_-}$$

See also **Electrophoretic mobility.**

Modified electrode. A thin layer of material (often an organic or organometallic polymer) coated on a conducting substrate such that direct electron transfer between substrate and electrolyte is not possible. The electrochemical reactions are between electrode and redox couples in the coat, and between the redox couple and electrolyte (*see* Figure 93). Typical coating

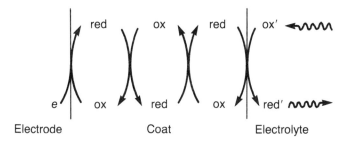

Figure 93. Electrochemical reaction at a modified electrode

198 Molar conductivity

materials are poly(vinyl pyridine), poly(vinyl ferrocene) and ruthenium 4,4′-bipyridyl. The coat may be attached to the substrate by direct chemical linkage via –O–Si–, adsorption, electrochemical coating or by plasma coating.

The theory of the action of modified electrodes is complex. Reaction may occur at the coat/electrolyte interface or somewhere inside the layer when either electron transfer from the electrode or reactant diffusing from the electrolyte is rate-limiting.

Applications of modified electrodes include oxygen reduction (*see* **Oxygen electrode reactions**), oxidations with Ru (IV) such as isopropanol to acetone, catalytic reduction of organic halides to alkenes, selective electrodes for photogalvanic cells and redox reactions of biologically important molecules.

Further reading
W. J. Albery and A. R. Hillman, Modified electrodes, in *Annual Reports C*, Royal Society of Chemistry, London, p. 377 (1981)

Molar conductivity (units: Ω^{-1} m^2 mol^{-1}). The **conductivity** of a solution at a concentration of 1 mol m^{-3}

$$\Lambda = \frac{\kappa}{c}$$

where

$$\kappa = \sum_i c_i \Lambda_i$$

whence

$$\Lambda = \frac{c_i(\Lambda_+ + \Lambda_-)}{c} \tag{86}$$

Similarly, the molal conductivity $\Lambda_m = \kappa/m$. The molar conductivity, unlike the conductivity, is not directly dependent on the concentration, but it does depend on two factors which may vary with concentration. For an incompletely dissociated electrolyte, the value of the molar conductivity depends on the ratio c_i/c (i.e. α, the degree of dissociation), thus

$$\Lambda = \alpha(\Lambda_+ + \Lambda_-) \tag{87}$$

It also depends on the variation of the **molar ionic conductivity** values of the ions with concentration. For a completely dissociated electrolyte, this is the only effect so that Eq. (87) becomes

$$\Lambda = \Lambda_+ + \Lambda_- \tag{88}$$

Electrolytic dissociation becomes more complete as the concentration is reduced, and at the limit of infinite dilution Eq. (88) becomes

$$\Lambda^\circ = \Lambda^\circ_+ + \Lambda^\circ_-$$

These limiting ionic conductivities correspond to the ideal state in which the ions are so far separated as to have no mutual influence.

In quoting the molar conductivity or the limiting molar conductivity for an electrolyte, the formula unit to which the value applies must be specified. When comparing values of the molar conductivity, it is essential to define 1 mole in all cases as the amount associated with 1 mole of unit charge. For 1:1 electrolytes there is no problem, but for electrolytes such as magnesium sulphate a mole would be specified as $\frac{1}{2}MgSO_4$ rather than $MgSO_4$. Thus $\Lambda^\circ_{MgSO_4} = 266.14 \times 10^{-4}\,\Omega^{-1}\,m^2\,mol^{-1}$, but $\Lambda^\circ_{\frac{1}{2}(MgSO_4)} = 133.07 \times 10^{-4}\,\Omega^{-1}\,m^2\,mol^{-1}$.

In the application of Kohlrausch's law, the values of the molar ionic conductivities must refer to the quantity of ions present in the specified amount of electrolyte. Thus

$$\Lambda^\circ_{\frac{1}{2}(MgCl_2)} = \tfrac{1}{2}\Lambda^\circ_{Mg^{2+}} + \Lambda^\circ_{Cl^-}$$
$$= 129.39\,\Omega^{-1}\,m^2\,mol^{-1}$$

It is equally correct to write

$$\Lambda^\circ_{MgCl_2} = \Lambda^\circ_{Mg^{2+}} + 2\Lambda^\circ_{Cl^-}$$
$$= 258.78\,\Omega^{-1}\,m^2\,mol^{-1}$$

A more general form of the Kohlrausch law is thus

$$\Lambda_{electrolyte} = z_+\nu_+\Lambda_+ + z_-\nu_-\Lambda_-$$

Prior to about 1968, the literature employed the quantity equivalent conductance, $\Lambda_{equiv} = 1000\,\kappa/c_{equiv}$ (units, $\Omega^{-1}\,cm^2\,g\,equiv^{-1}$), in which the concentration was expressed as equivalents per litre. To convert Λ_{equiv} to Λ, divide by 10^4 and multiply by the number of charges associated with the ionized molecule (i.e. $z\nu$), thus for aluminium sulphate $z\nu = 6$.

Molar ionic conductivity (Λ_i; units: $\Omega^{-1}\,m^2\,mol^{-1}$). Kohlrausch observed that the differences between the molar conductivities at zero concentration of several sodium and potassium salts were constant and concluded that each ion contributes to the total conductance of the electrolyte independently of the other ions present. Thus

$$\Lambda^\circ_{electrolyte} = z_+\nu_+\Lambda_+ + z_-\nu_-\Lambda_-$$

The molar ionic conductivity is the contribution of the ith ion to the total molar conductivity of the electrolyte. It is obtained by combining experimental values of the **molar conductivity** of the electrolyte with the **transport number** of each of the ions, thus

$$\Lambda_+ = t_+\Lambda \quad \text{and} \quad \Lambda_- = t_-\Lambda$$

200 Molar ionic conductivity

Typical values of the limiting values of the molar ionic conductivity of a range of ions in different solvents are tabulated (see Appendix, Table 12). In comparing the values for the different ions, care must be taken to ensure that the correct comparisons are made. Thus there is no significance in comparing Λ^o_{Na+}, Λ^o_{Zn2+} and Λ^o_{Al3+}, whereas there is real significance in comparing Λ^o_{Na+}, $\frac{1}{2}\Lambda^o_{Zn2+}$ and $\frac{1}{3}\Lambda^o_{Al3+}$ because the molar values, so defined, are referred in each case to the amount of material associated with one mole of electrons.

The H_3O^+ and OH^- ions have relatively very high values for the limiting molar ionic conductivity, and this is due to the strong tendency to hydrogen bonding between water molecules. As a result, the proton is able to move in the direction of the current by an exchange of partners much more rapidly than by a normal process of migration. The proton jump mechanism (Grotthus theory) for water is

A similar mechanism will apply to the hydroxide ion. (It is not necessarily the same proton which moves in a subsequent exchange.) Other ions cannot participate with the water molecule in this way.

Proton transfer under an applied field is not solely confined to aqueous systems: for example, the high conductivity of the HSO_4^- ion in concentrated sulphuric acid. The high cationic conductivity in anhydrous methanolic hydrogen chloride is due to a similar proton jump mechanism

$$CH_3\overset{+}{O}H_2 + CH_3OH \rightleftharpoons CH_3OH + CH_3\overset{+}{O}H_2$$

However, the presence of a trace of water will result in the irreversible transfer of the protons to water molecules.

The average limiting ionic conductivity for a range of monovalent ions in water is about 75×10^{-4} Ω^{-1} m^2 mol^{-1} (for potassium, rubidium, caesium, thallium, ammonium, chloride, bromide, iodide, nitrate, chlorate and perchlorate ions). These ions lie in a critical range; if smaller (in terms of crystallographic radius), they would acquire a permanent hydration sheath and end up larger with a lower limiting ionic conductivity (e.g., sodium, lithium and fluoride ions). If larger, they would hydrate but would be slower moving (i.e. a lower ionic conductivity) on account of their larger size.

For the alkali metal ions, the values of the limiting ionic conductivity are in the inverse order of their crystallographic radii. This is attributed to 'hydration' effects; the small lithium ion orientates and holds water molecules on account of the strong ion–dipole attraction, whereas the weaker attraction of the larger ions leads to a net 'structure-breaking' effect through its disruptive effect on the hydrogen bonding of the water molecules in its vicinity.

The tetraalkylammonium ions (NR_4^+) are of great interest since they combine large size and symmetrical shape with low charge and some of their salts are soluble in solvents other than water. The values of their limiting ionic conductance decrease with increasing chain length. For such ions, the product of $\Lambda_i^\circ \eta$ is constant over a range of temperatures.

The Stokes law radius (r_s) can be obtained from values of Λ_+°

$$r_s = \frac{0.820|z_+|}{\Lambda_+^\circ \eta_0}$$

Approximate values for the radius of such ions, obtained from molecular models or from molecular volumes, leads to a Stokes law correction (r/r_s) for the value calculated from the limiting ionic conductivity. Thus the true radius of any other hydrated ion may be calculated from

$$r = \frac{0.820|z_+|}{\Lambda_+^\circ \eta_0} \left(\frac{r}{r_s}\right)$$

leading to a value for the hydration number of the ion. Typical hydration numbers are: sodium, 5; lithium, 7; magnesium, 12; calcium, 10.

In all non-aqueous solvents, except sulpholane, the molar ionic conductivities of the alkali metal ions increase with increases in crystallographic radius i.e. lithium < sodium < potassium < rubidium < caesium (compare values in water). On the other hand, the values for NR_4^+ decrease as the size of R increases. The caesium and chloride ions, both with inert gas electronic structures and approximately equal radii, are used to test the effect of charge type; the caesium ion is the faster ion in protic solvents (except formamide and hydrogen cyanide), while the chloride ion is faster in aprotic solvents.

Temperature effects (Walden's law). Molar and molar ionic conductivities have large temperature coefficients. Since the conductance of an electrolyte depends on the mobility of the ions, the molar conductivity at zero concentration should be inversely proportional to the viscosity of the solvent at constant temperature, thus

$$\Lambda_i^\circ \eta_0 = \text{constant}$$

At constant temperature, the product $\Lambda_i^\circ \eta_0$ for a given ion varies from solvent to solvent and attains constancy only when the ion is very large. The constancy is poor for the picrate ion and NEt_4^+, and fair for NBu_4^+ (except in water, sulpholane and glycol). The larger symmetrical ions, $N(iso\text{-}Am)_4^+$ and BPh_4^- fare better, but the range of solvents is limited.

202 Molten salt systems

The product $\Lambda_i^\circ \eta_0$ for NR_4^+ and solvated ions such as lithium and lanthanum in water is constant over a wide range of temperatures, suggesting that the effective radius of these ions is unchanged with temperature. For the larger univalent ions ammonium, potassium, chloride and bromide, the product decreases by about 30 percent for an increase of temperature from 273 to 373 K. This suggests that the effective radius varies with temperatures and provides evidence for the solvation of ions in solution. See Covington and Dickinson (1973), chapter 5; Robinson and Stokes (1970).

Molten salt systems. *See* **Fused salt electrochemistry; Fused salts.**

Moving boundary method. *See* **Isotachophoresis; Transport number.**

N

Nafion membrane. *See* **Ion-selective membrane.**

Nernst diffusion layer. *See* **Diffusion-limited current.**

Nernst equation. An equation relating the **electrode potential** of a reversible electrode to the activity of the ions in solution.

When the **half cell** (electrode) is coupled with a standard hydrogen electrode to give a cell

$$\text{Pt,H}_2(g) \mid \text{H}^+ \mid \text{KCl} \mid \text{M}^{z+} \mid \text{M}$$
$$(1 \text{ atmos}) \quad (a_{H3O+} = 1)$$

at equilibrium the sum of the **electrochemical potential** on each side of each electrode reaction must balance, that is

$$\sum_i \nu_i \bar{\mu}_i = 0$$

For the left-hand half cell, the electrode reaction is

$$\text{H}_3\text{O}^+ + e \rightleftharpoons \tfrac{1}{2}\text{H}_2 + \text{H}_2\text{O}$$
$$(a_{H3O+} = 1) \quad (1 \text{ atmos})$$

for which

$$\bar{\mu}_{H3O+} + \bar{\mu}_e = \tfrac{1}{2}\mu_{H2} + \bar{\mu}_{H2O}$$

From the definition of the **electrochemical potential** (*see* Eq. (56))

$$(\mu^\ominus_{H3O+} + RT \ln a_{H3O+} + F\phi_{ion}) + (\mu_e - F\phi_e) = \tfrac{1}{2}(\mu^\ominus_{H2} + RT \ln p_{H2}) + \mu_{H2O}$$

(since hydrogen and water are uncharged, $z_i = 0$), thus on rearranging

$$F(\phi_e - \phi_{ion}) = \mu^\ominus_{H3O+} + \mu_e - \tfrac{1}{2}\mu^\ominus_{H2} - \mu_{H2O} + RT \ln \frac{a_{H3O+}}{p_{H2}^{1/2}}$$

$$= A + RT \ln \frac{a_{H3O+}}{p_{H2}^{1/2}} \tag{89}$$

where A is a constant and $\phi_e - \phi_{ion}$ represents the potential of the electron donated to the platinum electrode relative to the potential of the ion in solution, thus Eq. (89) becomes

204 Nernst equation

$$\phi_{Pt} - \phi_{soln} = A + \frac{RT}{F} \ln \frac{a_{H3O+}}{p_{H2}^{1/2}} \tag{90}$$

Similarly for the right-hand half cell

$$\frac{1}{z} M^{z+} + e \rightleftharpoons \frac{1}{z} M$$

at equilibrium this becomes

$$\frac{1}{z} \bar{\mu}_{Mz+} + \bar{\mu}_e = \frac{1}{z} \bar{\mu}_M$$

which on substitution of the electrochemical potential (Eq. (56)) and rearranging gives

$$F(\phi_e - \phi_{ion}) = \left(\frac{1}{z} \mu^{\ominus}_{Mz+} + \mu_e - \frac{1}{z} \mu_M \right) + \frac{RT}{z} \ln a_{Mz+} \tag{91}$$

ϕ_e is now the potential of the metal and the chemical potentials in parenthesis are constant, thus Eq. (91) becomes

$$\phi_M - \phi_{soln} = B + \frac{RT}{zF} \ln a_{Mz+} \tag{92}$$

Subtracting Eq. (90) from Eq. (91) gives the e.m.f. of the cell

$$E = (\phi_M - \phi_{soln}) - (\phi_{Pt} - \phi_{soln}) = \phi_M - \phi_{Pt}$$

$$= (B - A) + \frac{RT}{zF} \ln \frac{a_{Mz+}}{(a_{H3O+}/p_{H2}^{1/2})^z} \tag{93}$$

$(B - A) = E^{\ominus}_{Mz+}$, the standard electrode potential of the half cell $M^{z+}|M$ since by convention the standard hydrogen electrode potential is zero, thus when $a_{H3O+} = 1$ and $p_{H2} = 1$ atmos, Eq. (93) becomes

$$E = E^{\ominus}_{Mz+,M} + \frac{RT}{zF} \ln a_{Mz+} \tag{94}$$

Eq. (94) is the traditional Nernst equation for a cationic electrode but expressed in terms of the e.m.f. of the cell studied.

A second approach considers the calculation of the free energy change of the cell reaction which is assumed to go to completion; using the relationship

$$\Delta G = \sum_i \nu_i \mu_i = -nFE \tag{95}$$

The cell reaction for the above cell (obtained by summing the separate half-cell reactions) is

$$\tfrac{1}{2} H_2 + H_2O + \frac{1}{z} M^{z+} \rightleftharpoons H_3O^+ + \frac{1}{z} M$$

for which $n = 1$. Thus

$$\Delta G = \mu_{H3O+} + \frac{1}{z}\mu_M - \frac{1}{2}\mu_{H2} - \mu_{H2O} - \frac{1}{z}\mu_{Mz+} \quad (96)$$

Substituting the values of the **chemical potential** (Eq. (15)) for the various species and combining Eq. (95) and Eq. (96) gives

$$E = E^\ominus_{Mz+,M} - \frac{RT}{zF} \ln \frac{a_{H3O+}}{a_{Mz+}} \quad (97)$$

Since $a_{H3O+} = 1$ for the standard hydrogen electrode, Eq. (97) can be rearranged to give the Nernst equation (Eq. (94)).

For electrodes reversible to anions (e.g., the chlorine electrode)

$$\tfrac{1}{2}X_2 + ze \rightleftharpoons X^{z-}$$

the Nernst equation becomes

$$E = E^\ominus_{Xz-,X2} - \frac{RT}{zF} \ln \frac{a_{Xz-}}{p^{1/2}_{X2}}$$

In a similar manner, the Nernst equation for the redox cell

$$\begin{array}{c|c|c|c} Pt, H_2 & H^+ & KCl & ox, red \mid Pt \\ (1 \text{ atmos}) & (a_{H3O+} = 1) & & \end{array}$$

in which the half-cell reaction is

$$ox + ne \rightleftharpoons red$$

is

$$E = E^\ominus_{ox,red} + \frac{RT}{nF} \ln \frac{a_{ox}}{a_{red}} \quad (98)$$

where $E^\ominus_{ox,red}$ is the standard redox potential. This is the general form of the Nernst equation, for cationic systems the oxidized form is M^{z+}, and the reduced form M, whereas for the anionic systems the oxidized form is X_2 and the reduced form X^- (X, for example, could be chlorine).

Nickel–cadmium battery. A mechanically rugged, long-lived secondary battery which has electrodes of nickel hydroxide and cadmium hydroxide, with potassium hydroxide as electrolyte. The cell reactions are

cathode $\quad 2NiO(OH) + 2H^+ + 2e \rightleftharpoons 2Ni(OH)_2$
anode $\quad\quad\quad Cd + 2H_2O \rightleftharpoons Cd(OH)_2 + 2H^+ + 2e$
---
overall $\quad Cd + 2NiO(OH) + 2H_2O \rightleftharpoons 2Ni(OH)_2 + Cd(OH)_2$

This battery may be sealed, in which case the anode contains excess cadmium hydroxide so that if the battery is overcharged only oxygen is evolved at the nickel hydroxide electrode. The oxygen is allowed to recombine at the anode

$$\tfrac{1}{2}O_2 + Cd + H_2O \rightarrow Cd(OH)_2$$

The cost of a nickel–cadmium battery is higher than that of the **lead–acid battery** or nickel–zinc battery.

A nickel–cadmium battery may be fashioned in the form of a **button cell** giving 1.2 V, 0.1 A h at 10 W up to 1 A continuous discharge. See Crompton (1982).

Nickel electrometallurgy. A major proportion of nickel produced is now obtained electrolytically. The concentrate reaches the electrolysis stage as matte, cast in the form of anodes. These consist mainly of the sulphide (Ni_3S_2) together with metallic nickel; copper, iron and cobalt are the most important impurities. In the electrolytic tanks, the nickel cathodes are separated from the matte anodes by diaphragms of synthetic cloth. The electrolyte, a solution of nickel sulphate and nickel chloride, enters the cathode compartment, overflows from the anode compartment and is chemically treated before recirculation. Electrolysis requires a cell voltage of 3–4 V. The whole of the sulphur remains in the anode sludge together with some precious metal residues, from which it is separated by melting and filtration.

The current efficiency at the matte anodes is only about 95 percent, so that a deficiency of nickel and an increase in acidity develop in the electrolyte. The chemical treatment of the latter must therefore correct for this as well as remove impurities. The electrolyte at pH 1.5 is first treated with hydrogen sulphide, which precipitates copper and arsenic (III) sulphides; copper is recovered from the precipitate. The electrolyte then flows to a series of agitated aeration tanks and finely ground anode scrap is added. Metallic nickel dissolves and the sulphide undergoes the reaction

$$Ni_3S_2 + \tfrac{1}{2}O_2 + 2H^+ \rightarrow Ni^{2+} + 2NiS + H_2O$$

The filtered solution is next treated with chlorine to oxidize the iron and cobalt; these are precipitated at a controlled pH by the addition of nickel carbonate. Cobalt is removed from the precipitate, and the filtrate is now ready for return to the electrolytic tanks.

Non-aqueous solutions. For electrochemical purposes, non-aqueous solvents have the potential advantage of enabling processes to be carried out which are impossible in water. Such solvents frequently have a much lower conductance than water with the result that measurements can be carried out at much lower concentrations without serious loss of accuracy. On the other hand, they are more difficult to purify and usually require careful protection from atmospheric moisture, whereas simple salts are often only slightly soluble with a consequent reduction in the concentration range that can be studied.

Solvents are divided into two main groups—those with relative permittivity below and above 30—and each group is further subdivided according to whether they are hydrogen-bonded or not (*see* Table 25). The solvent influences the conductance of a salt through three main factors: (1) its viscosity, (2) its relative permittivity and (3) its interaction with the ions (solvation), which will be determined by the structure, and the size of the solvent molecule and the location and orientation of any dipole(s) it contains.

The molar conductivity at infinite dilution is governed primarily by the viscosity; thus the molar conductivities of the common ions are much higher in acetone than they are in water. Where data are lacking, approximate values for Λ° can be calculated from those for another solvent by using Walden's equation

$$\Lambda^\circ \eta = \text{constant}$$

This equation ignores solvent/solute interaction, but is fairly reliable for large ions, such as the tetraalkylammonium ions, and for the more non-polar solvents, where solvation will not be expected to be critically important.

Measurements at very low concentrations have confirmed the validity of Onsager's equation (*see* **Conductance equations**) for a large number of systems, and it can be assumed to hold generally. At finite concentrations, however, the relative permittivity of the solvent becomes the most important factor. Values for some of the solvents which have been employed (*see* Appendix, Table 2) are generally considerably lower than that of water. In these, coulomb attractions are far stronger, the Onsager slopes are much steeper, and ion pairing is much more general and more extensive than in water, so that the conductance is liable to fall off rapidly to low values.

The straight line (*see* Figure 94) obtained by plotting the pK of tetraethylammonium picrate, measured in a range of solvents, against the reciprocal of the relative permittivity is that which would be expected if ion pairing was governed purely by electrostatic forces. For these two large ions, the picture does not appear to be complicated by solvation effects or by special (non-coulombic) attractive forces between the ions.

The wide scatter of the points for lithium picrate is attributed to solvation. Four of the points are approximately on a straight line, and in the solvents concerned (acetonitrile, acetone, methyl ethyl ketone and pyridine) it is probable that a small cation will be strongly solvated. The other two points are for solutions in nitromethane and nitrobenzene, solvents with comparatively high values for the relative permittivity. Solvation appears too weak to protect the ions from ion association and in consequence the salts are extremely weak.

Table 25
General classification of solvents

$\varepsilon_r > 30$		$\varepsilon_r < 30$	
Hydrogen-bonded	Non-hydrogen-bonded (dipolar aprotic)	Hydrogen-bonded	Non-hydrogen-bonded
Water	Acetonitrile	Ethanol	Acetic anhydride
Methanol	Dimethylacetamide	iso-Propanol	Acetone
Ethylene glycol	Dimethylformamide	tert-Butanol	Benzene
Anhydrous sulphuric acid	Dimethylsulphoxide	Glacial acetic acid	Carbon tetrachloride
Hydrogen fluoride	Nitrobenzene	Trichloroacetic acid	Chlorobenzene
Formic acid	Nitromethane	Liquid ammonia	Chloroform
Hydrocyanic acid	Tetramethylene sulphone (sulpholane)	Ethylene diamine	Cyclohexane
Formamide	Propylene carbonate		Dioxan
Ethanolamine			Liquid sulphur dioxide
Acetamide			Diethyl ether
N-Methyl formamide			Pyridine
N-Methyl acetamide			

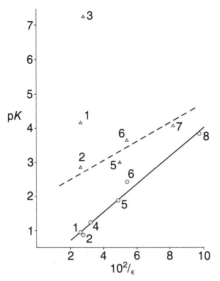

Figure 94. pK values in non-aqueous solvents: ○ tetraethylammonium picrate; △ lithium picrate. Solvents: 1, nitromethane; 2, acetonitrile; 3, nitrobenzene; 4, methanol; 5, acetone; 6, methyl ethyl ketone; 7, pyridine; 8, ethylene chloride

The limiting **molar ionic conductivity** values of both anions and cations increase as the crystallographic radius increases (cf. the corresponding values for water where there is no relationship between the molar ionic conductivity and the radius).

The standard **electrode potential** is often not greatly different in non-aqueous solvents from that in water, although displacements due to differences in the strength of solvation of the ions are to be expected. The same reference electrodes as are used in water are also usually satisfactory, especially those based on sparingly soluble salts as they can be combined with other electrodes in cells without a liquid junction. *See also* **Acids and bases; Conductance minima; Conductance of non-aqueous solutions; Transport number;** Appendix, Table 12. See Covington and Dickinson (1973); Ives and Janz (1961); Janz and Tomkins (1972); Janz and Tomkins (1973).

Further reading
J. Barthel, R. Wachter and H. J. Gores, Temperature dependence of conductance electrolytes in nonaqueous solutions, in *Modern Aspects of Electrochemistry*, vol. 13, B. E. Conway and J. O'M. Bockris (ed.), Plenum Press, New York, p. 1 (1979)

O

Ohm (Ω; dimensions: $\varepsilon^{-1} \, l^{-1} \, t$; units: kg m^2 s^{-3} A^{-2} = V A^{-1}). Unit of electrical resistance through which a potential difference of 1 volt will produce a current of 1 ampere, thus

$$E = IR$$

The international ohm is the resistance offered to an unvarying current by a column of mercury at 0°C, 14.452 1 g in mass, of constant cross-sectional area and 106.300 cm in length. *See also* **Electric units**.

Ohm's law. The current passing through a conductor is directly proportional to the applied e.m.f.; that is

$$E = IR$$

In addition to electronic conductors, electrolytic solutions, in the absence of complicating factors such as electrolysis, obey Ohm's law. *See also* **Conductivity**.

Onsager's equation. *See* **Conductance equations**.

Open-circuit voltage. When no current flows in an electrochemical cell the potential of the working electrode against the reference electrode is known as the open circuit voltage (OCV) or rest potential. In cells in which all the species that contribute to the overall electrochemical reaction are present, the OCV is given by the **Nernst equation**. In cases such as gas-evolution reactions, the OCV may depend on some other electrochemical couple: for example, the OCV of oxygen evolution at platinum is at a potential which probably corresponds to a PtO|PtO$_2$ half cell.

Optically transparent electrode (OTE). A micromesh or thin film electrode through which light may pass. Very thin layers of semiconductors (e.g., stannic oxide) or metals (e.g., platinum, gold or silver) may be deposited on transparent substrates such as glass, quartz or certain plastics by vapour-phase deposition, from a radio frequency discharge or plasma, or by reduction of a solution. A wider range of materials may be deposited on a gold micromesh electrode which acts as a planar electrode if the diffusion layer thickness (*see* **Diffusion-limited current**) is greater than the mesh size.

Organic electrochemistry

In cases in which a thin layer of solution is also required, for example in the spectroscopic analysis of an exhaustively electrolysed solution, an optically transparent thin layer electrode (OTTLE) is constructed consisting of a thin sandwich of microscope slides which support the OTE. Electrolyte from the cell reservoir is drawn up between the slides by capillary action. The reference electrodes and counterelectrodes are also located in the reservoir. *See also* **Spectroelectrochemistry**.

Organic electrochemistry. This subject dates back to the earliest research in electrochemistry by Faraday, Nernst and Haber. The advantages of using an electrode potential to effect an organic reaction are:

1) Precise control of the potential allows the possibility of highly selective reactions.
2) An elevated temperature is not required.
3) Added reagents, such as oxidants and reductants, which may interfere with the products of the reaction may be avoided.
4) Reactions may be followed and controlled very accurately by monitoring the current and voltage.

Disadvantages are based on the resistivity of solutions, and the low conversion rates of many organic electrochemical reactions.

Organic electrochemistry is used for synthesis and to investigate reaction mechanisms. Some important industrial syntheses are described in **organic synthesis**, but it must be borne in mind that if the products of any reaction are useful and are obtained in sufficient yield, then that reaction is a possible synthetic route to the product.

Apparatus and techniques. The design of a cell will depend: on the complexity of the reaction, for example whether a fixed voltage against a reference is required; if the products of the anodic and cathodic reactions must be separated; if the products or reactants are air-sensitive; and so on. One criterion which is usually important is the need to have a large electrode surface area to volume ratio. This minimizes ohmic losses in the electrolyte, and, for micropreparative work, allows small amounts of reactants to be used.

Separation in a cell is effected by means of a cell divider which may be of ceramic material (porous pot, sintered glass) or organic polymer (cellophane membrane, PTFE, Nafion, Flemion). Ceramic dividers show good stability in all electrolytes, but are not in any way selective, allowing passage of solvent and all ions. Organic membranes, especially an **ion-selective membrane**, are selective, but are not so robust as ceramic dividers, and these may introduce large ohmic losses.

The choice of electrode material governs the potential range over which the electrochemical reactions may be performed, and the degree of **electrocatalysis** which may occur. The limitation of the cathodic reaction in aqueous media is the onset of the evolution of hydrogen. Metals with low exchange current densities for the hydrogen evolution reaction (*see* **Hydrogen electrode reactions**) (e.g., lead and mercury) are to be preferred to highly active noble metals (e.g., platinum, palladium and rhodium). In non-aqueous media, the cathodic limit is set by the choice of solvent. The anodic limit is governed by the oxidation of the electrode material, and, in aqueous solution, the oxidation of water. Lead and mercury are easily oxidized and so may not be used as anodes. Lead dioxide, however, is a powerful oxidizing agent and is stable at high anodic potentials. Platinum, although oxidized in aqueous electrolyte, is sufficiently stable to be used as an anode to the potential at which oxygen is evolved at this metal (approximately 1.7 V versus standard hydrogen electrode).

In non-aqueous media, a problem arises in making contact between an aqueous reference electrode (e.g., saturated **calomel electrode**) and the solution. A junction potential of 30–40 mV arises which must be kept constant. Aqueous reference electrodes may not be used if the organic solution is water-sensitive. These problems may be resolved by the use of a non-aqueous reference electrode: for example, calomel in dimethylformamide or methanol with the potassium chloride electrolyte replaced by tetraethylammonium chloride has been reported. The solvent and electrolyte are chosen for their stability to the potential range of the experiment, the solubility of the starting material and the conductivity of the solution. Typical solvents are acetonitrile ($+2.4$ to -3.5 V versus saturated calomel electrode), dimethylformamide ($+1.6$ to -3.0 V versus saturated calomel electrode), hexamethylphosphoramide ($+0.75$ V versus $Ag|Ag^+$), pyridine ($+1.4$ to -2.3 V versus $Ag|Ag^+$), tetrahydrofuran ($+1.5$ to -4 V versus $Ag|Ag^+$) and alcohols (approximately 0 to -2 V versus saturated calomel electrode). Aqueous saturated calomel electrodes may be used with all these solvents. The electrolyte must dissolve and ionize in the chosen solvent. Tetraalkylammonium salts have been used extensively to provide cations. Perchlorates and halide ions are common anions.

Reduction of organic compounds. Nitro compounds are reduced to amines via nitroso compounds and hydroxylamines: for example, nitrobenzene is reduced to aniline in a six-electron reduction in acid media at a lead or mercury cathode.

$$PhNO_2 + 2H^+ + 2e \rightarrow PhNO + H_2O$$
$$PhNO + 2H^+ + 2e \rightarrow PhNHOH$$
$$PhNHOH + 2H^+ + 2e \rightarrow PhNH_2 + H_2O$$

The importance of the conditions is seen in this reaction. In basic solution, a bimolecular reaction occurs giving azobenzene, while at a platinum cathode the intermediate hydroxylamine rearranges to *p*-aminophenol. In aqueous hydrochloric acid, *o*- and *p*-chloroaniline are formed. The first two steps are thought to proceed quickly by direct electron transfer. The final stage may be a reduction by atomic hydrogen.

Carbonyl groups are reduced to the corresponding alcohol in aqueous electrolyte in two one-electron reactions

$$R_2CO + H^+ + e \rightarrow R_2C\dot{\,}OH$$
$$R_2C\dot{\,}OH + H^+ + e \rightarrow R_2CHOH$$

Further reduction to an alkane is possible

$$R_2CHOH + 2H^+ + 2e \rightarrow R_2CH_2 + 2H_2O$$

The intermediate radical $R_2C\dot{\,}OH$ may dimerize to give the pinacol $(R_2COH)_2$. The route chosen by the reaction is dependent on the electrode material. Acetone is reduced to pinacol and isopropanol at lead, but isopropanol and propane at mercury.

Alkyl halides are reduced to a radical and halide ion. This radical may undergo further reduction to an alkane or it may dimerize. An important use of this reaction is in the production of tetraethyl lead (*see* **Organic synthesis**). A wide variety of organic syntheses are possible by the electrolytic production of a reactive species (radical or radical ion) in a suitable solution of the necessary reactants.

Oxidation of organic compounds. As a synthetic method, electrooxidation has proved less useful than electrolytic reduction. The oxidation of an aliphatic hydrocarbon through the steps

hydrocarbon → alcohol → aldehyde → acid → $CO_2 + H_2O$

may give desirable products at a particular stage, but the control necessary to give a good yield is difficult to achieve and well-tried chemical methods often provide a satisfactory alternative. The scheme hydrocarbon → $CO_2 + H_2O$ is, however, the route by which the energy of an exoenergetic organic compound may be released electrochemically in a **fuel cell**: for example, methane may be electrochemically oxidized

$$CH_4 + 2H_2O \rightarrow CO_2 + 8H^+ + 8e$$

Coupled with an air cathode, the overall reaction, which yields 1.06 V, is

$$CH_4 + 2O_2 \rightarrow CO_2 + 2H_2O$$

Oxidations of major synthetic importance are the Kolbe reaction and anodic fluorination (*see* **Organic synthesis**). Characteristic of aromatic compounds is the anodic oxidation of benzene which can yield the following products

214 Organic synthesis

Benzene → Phenol → Catechol (OH, OH)

Phenol → Hydroquinone (HO, OH) → Quinone (O=, =O) → CO₂H CH = CH CO₂H (Maleic acid)

Benzene →(I₂, Pt)→ Iodobenzene

Anodic chlorination involves the production *in situ* of chlorine by oxidation of a reactant, followed by a chemical chlorination. Thus ethanol is converted to chloroform on oxidation in an aqueous solution of sodium chloride by HOCl, which is generated by the dissolution of chlorine. Iodoform is prepared electrochemically in a similar manner.

Further reading
Acta. chem. scand., **B37**, no. 5–6, (1983)
M. M. Baizer, *Organic Electrochemistry*, Marcel Dekker, New York (1973)
D. H. Morman and G. A. Harlow, Organic reference electrodes, *Anal. Chem.*, **39**, 1869 (1967)
M. R. Rifi and F. H. Covitz, *Introduction to Organic Electrochemistry*, Marcel Dekker, New York (1974)

Organic synthesis. Electrosynthetic organic chemistry started in 1801 with the electrochemical oxidation of ethanol by Erman. Michael Faraday investigated the electrochemistry of alcohols, acids and salts, and in 1849, Kolbe published his work on the electrosynthesis of alkanes by the oxidation of carboxylic acids.

Although electroorganic syntheses may mirror chemical syntheses, there are many reactions that are unique to electrochemistry: for example, those that rely on the ease of electrochemical production of radical ions. Mechanisms and techniques are dealt with under **organic electrochemistry**; here we shall concentrate on important synthetic processes.

Kolbe synthesis. Electrolysis of a solution of an alkali metal salt of a carboxylic acid yields an alkane and carbon dioxide

$$RCO_2^- + R'CO_2^- \rightarrow R\text{-}R' + 2CO_2 + 2e$$

The Kolbe reaction is favoured by high salt concentrations, high current density and a platinum anode. Under nonideal conditions, a variety of side products—alkenes, esters, rearranged alkanes—are formed. These are produced by reactions of an intermediate radical, or carbonium ion which is the result of further oxidation of the radical. Reactions via a carbonium ion are known as two-electron pathways. Kolbe reactions have also been used as initiators for polymerizations. Industrially the Kolbe synthesis is used to manufacture dimethyl sebacate (a precursor of sebacic acid) from sodium monomethyl adipate

$$2CH_3OOC(CH_2)_4COONa$$
$$\rightarrow CH_3OOC(CH_2)_8COOCH_3 + 2Na^+ + 2CO_2 + 2e$$

Adiponitrile synthesis. The synthesis of adiponitrile, an intermediate in the manufacture of nylon 66, is by the electrohydrodimerization of acrylonitrile

$$2CH_2CHCN + 2H_2O + 2e \rightarrow CN(CH_2)_4CN + 2OH^-$$

The anode reaction is the oxidation of water. The commercial use of this electrochemical reaction arrived with the discovery by Monsanto of the importance of the supporting electrolyte to the selectivity of the process. Quantitative yields of adiponitrile are produced at lead or mercury cathodes with certain quaternary ammonium salts, such as tetraethylammonium *p*-toluenesulphonate, as electrolyte. A catholyte, consisting of an aqueous solution of acrylonitrile, the quaternary ammonium salt and reaction products, is circulated through the cathode compartments of a series of cells. Sulphuric acid is the anolyte; an ion-exchange membrane separates the two compartments (*see* **Ion-selective membrane**).

Tetraalkyl lead production. Tetraethyl or tetramethyl lead is a major additive to petrol as an antiknock agent. A Grignard reagent is electrochemically prepared with an excess of alkyl chloride. For example, to prepare tetraethyl lead the reaction at a sacrificial lead anode is

$$4C_2H_5^- + Pb \rightarrow (C_2H_5)_4Pb + 4e$$

and at the cathode

$$4MgCl^+ + 4e \rightarrow 2Mg + 2MgCl_2$$

In conditions of excess ethyl chloride, the magnesium formed at the cathode reacts to regenerate the Grignard reagent

$$Mg + C_2H_5Cl \rightarrow MgCl^+C_2H_5^-$$

The overall reaction is

$$2C_2H_5MgCl + 2C_2H_5Cl + Pb \rightarrow (C_2H_5)_4Pb + 2MgCl_2$$

The cell design uses lead pellets as the anode separated from the steel cathode which also acts as the walls of the tube cell and as a heat exchanger.

Electrofluorination. A hydrocarbon dissolved in liquid hydrogen fluoride is electrolysed at a nickel anode. Cathodes of nickel, iron, copper or platinum are suitable for the evolution of hydrogen. Sodium or potassium fluoride is added to increase the conductivity of the solution. Water must be excluded as the formation of oxygen difluoride creates an explosive mixture with the organic compounds. The extent of fluorination of the hydrocarbon is determined by the feed rate and current.

Electroinitiation of polymerization. Polymerization may be induced electrolytically in a number of ways. Discharge of an ion at an electrode may produce an atom, radical or radical ion which can initiate polymerization. A stable polymerization initiator may be produced by electrolysis, or an inhibitor may be removed, thus allowing polymerization to take place. The advantages of electrochemical initiation are: (1) the purity of the process (additives should be unnecessary); (2) the degree of polymerization may be controlled by the rate of electrolysis; (3) low temperatures are usually possible. An example is the polymerization of 1,3-butadiene

$$CH_2=CH-CH=CH_2 \xrightarrow[\text{initiation}]{+2e} \begin{bmatrix} \dot{C}H-CH_2 \\ | \\ CH=CH_2 \end{bmatrix}^{2-} \xrightarrow[\text{termination}]{-2e} \begin{bmatrix} CH-CH_2 \\ | \\ CH=CH_2 \end{bmatrix}_n$$

The current is pulsed through a cell, cathodically to form the anion, and anodically to terminate the polymerization. The number of pulses determines the amount of polymer, and the molecular weight is governed by the duration of the initiating and terminating pulses.

Electropolymerizations may be classed as cationic, anionic, free radical or condensation, depending on the nature of the intermediate involved.

Laboratory-scale syntheses. Many novel small-scale electroorganic syntheses have been discovered. These are reviewed in the Royal Society of Chemistry's series of *Specialist Periodical Reports* published annually.

Further reading

L. Meites and P. Zuman, *CRC Handbook Series in Organic Electrochemistry*, vol. I–IV, CRC Press, Cleveland (1978)

N. L. Weinberg (ed.), *Techniques of Chemistry*, vol. V, *Technique of Electroorganic Synthesis*, Wiley–Interscience, New York (1974)

Ostwald's dilution law. *See* **Arrhenius electrolytic dissociation theory.**

OTE. *See* **Optically transparent electrode.**

Outer Helmholtz plane. *See* **Electrical double layer.**

Outer potential (ψ). Between approximately 10^{-3} and 10^{-5} cm from a metal surface there exists a region in which the potential due to the metal is independent of distance. The outer, Volta or psi (ψ) potential is that potential near an electrified interface just outside the range of image forces. An outer potential may also be defined for an electrolyte and thus, for an electrode immersed in an electrolyte, the outer potential difference is given by

$$\Delta\psi = \psi_{metal} - \psi_{solution}$$

$\Delta\psi$ may be measured or calculated from simple electrostatics and a reasonable model of the structure of the metal/solution interface.

Overpotential (η). The potential (ϕ) of an electrode through which current flows differs from that at equilibrium (ϕ_e) by an amount known as the overpotential or overvoltage

$$\eta = \phi - \phi_e = E - E_e$$

where E refers to a measurement made against a reference electrode. η may be positive or negative depending on whether the deviation from equilibrium is anodic or cathodic. A number of factors contribute to the overpotential.

1) *Charge transfer or activation overpotential.* This occurs by virtue of the activation energy for the transfer of an electron during the redox reaction (*see* **Butler–Volmer equation**).
2) *Diffusion or concentration overpotential.* This arises from the rate-limiting mass transport of particles to or from the bulk to the electrode surface (*see* **Diffusion-limited current**).
3) *Reaction overpotential.* In a multistep electrochemical reaction, a rate-determining chemical reaction will contribute to the overpotential even though the reaction itself is independent of potential.
4) *Crystallization overpotential.* In metal deposition reactions, there is an activation barrier to the incorporation of a metal atom into the crystal lattice. This activation energy gives rise to the crystallization overpotential.

In addition, any ohmic resistances will contribute to the measured overpotential (*see* **Interruptor technique; Luggin capillary**).

Overvoltage. *See* **Overpotential.**

218 Oxidation

Oxidation. A process in which electrons are removed from a species, for example at the **anode** in a galvanic cell

$$\text{red}_1 \rightarrow \text{ox}_1 + ne$$

Reduction is a process in which electrons are donated to a species, for example, at the cathode in a galvanic cell

$$\text{ox}_2 + ne \rightarrow \text{red}_2$$

If the two processes are brought together in a cell, then electrons will be transferred from the anode to the cathode through the outside circuit and the overall reaction will be

$$\text{red}_1 + \text{ox}_2 \rightarrow \text{ox}_1 + \text{red}_2$$

Oxidation–reduction systems. *See* **Redox electrode systems.**

Oxide electrode. An example of an electrode of the second kind. The reversible electrode, represented by $OH^-|MO,M(aq)$, must conform with the equilibria

$$M^{2+} + 2e \rightleftharpoons M$$
$$MO(s) \rightleftharpoons M^{2+} + O^{2-}$$
$$O^{2-} + H_2O \rightleftharpoons 2OH^-$$
$$2OH^- + 2H^+ \rightleftharpoons 2H_2O$$

The potential of such an electrode is given by

$$E = E^{\ominus}_{M2+,M} + \frac{RT}{2F} \ln a_{M2+}$$

$$= E^{\ominus}_{M2+,M} + \frac{RT}{2F} \ln K_s - \frac{RT}{F} \ln K_w + \frac{RT}{F} \ln a_{H3O+}$$

where $K_s = a_{M2+} a^2_{OH-}$ and $K_w = a_{H+} a_{OH-}$ can be regarded as constant, thus

$$E = E' + \frac{RT}{F} \ln a_{H+}$$

$$= E' - \frac{2.303\, RT}{F} \text{pH}$$

The metal chosen must have a positive and reproducible electrode potential to resist attack, and the oxide must be stable, available in a reproducible state and sparingly soluble in water to avoid contamination of the test solution. The oxide electrode can be prepared by mixing the powdered metal with the powdered oxide, or by aerial or anodic oxidation. The **antimony–antimony oxide electrode** is the most common, but silver, mercury, tungsten, molybdenum and tellurium are among other metals used for oxide electrodes.

They are robust electrodes used for the control and measurement of pH; a few are capable of measurements of high accuracy. They find use in solutions of high alkalinity and at high temperatures, where a **glass electrode** and a **quinhydrone electrode** cannot be used, and as a replacement for the **hydrogen electrode** under conditions which are poisonous to that electrode. See Ives and Janz (1961), chapter 7.

Oxide film. A thin layer of oxide at the surface of a metal, whether formed naturally by exposure to the atmosphere or by anodic polarization. Oxide films have great effect on the properties of the surface and electrochemical reactions at the electrode. Examples of the mediation of the properties of metals by oxide films are:

1) Coherent layers of oxide may protect the underlying metal against corrosion by the process of **passivity**.
2) Conducting noble metal oxides have been found to be more efficient for chlorine evolution than the metals themselves (*see* **Dimensionally stable anode**).
3) Thin layers of oxides on semiconductors (e.g., silica) leads to important electronic devices in which electron tunnelling through the film is the dominant electron transfer mechanism.
4) Oxides of metals are frequently used as battery cathodes (*see* **Alkaline cell**), and may be recharged as in the **lead–acid battery.**
5) Oxides are thought to be an intermediate in the evolution of oxygen at many metals (*see* **Oxygen electrode reactions**).

Although most oxides of metals have known crystal structures, thin films of oxides can often be amorphous. In addition, the surface potential of an oxide film in contact with electrolyte is greatly dependent on the preferential adsorption of hydrogen or hydroxide ions. Schematically this may be represented as

$$-M-OH + H^+ \rightleftharpoons -M-OH_2^+$$

$$-M-OH + OH^- \rightleftharpoons -M-O^- + H_2O$$

$$-M-OH + OH^- \rightleftharpoons -M-(OH)_2^-$$

The double layer structure of surface oxides is characterized by high surface charge, but low diffuse layer charge (*see* **Electrical double layer**).

220 Oxygen electrode reactions

The uncertainty of the nature of oxide layers, especially when the metal may show more than one oxidation state, is a possible reason for the apparent breakdown of the **Nernst equation** for oxide half cells. An equation which relates the surface electrode potential to the pH and the concentration of positively (N_+) and negatively (N_-) charged surface sites for the metal–metal oxide half cell is

$$\psi = -2.303 \frac{RT}{F} (\text{pH} - \text{pH}_{\psi=0}) + \frac{RT}{2F} \ln \frac{N_-}{N_+}$$

where $\text{pH}_{\psi=0}$ is the pH at which the surface charge is zero (*see* **Potential of zero charge; Surface potential**).

Further reading
S. J. Morrison, *Electrochemistry at Semiconductor and Oxidized Metal Electrodes*, Plenum Press, New York (1980)
S. Trasatti (ed.), *Electrodes of Conductive Metal Oxides*, Elsevier, Amsterdam (1980)

Oxygen electrode reactions
Cathodic reaction. The reduction of oxygen is a most important reaction in **fuel cell** electrochemistry, it being the most common cathodic process

$$O_2 + 4H^+ + 4e \rightarrow 2H_2O \quad \text{in acid} \tag{a}$$

and

$$O_2 + 2H_2O + 4e \rightarrow 4OH^- \quad \text{in alkali} \tag{b}$$

The 'four-electron' routes given above are difficult to achieve. On many electrodes, a 'two-electron' process to hydrogen peroxide occurs

$$O_2 + 2H^+ + 2e \rightarrow H_2O_2 \tag{c}$$
$$O_2 + H_2O + 2e \rightarrow HO_2^- + OH^- \tag{d}$$

The subsequent reaction of hydrogen peroxide is slow and the existence of this competing mechanism leads to the highly irreversible nature of the oxygen electrode reaction. Experiments using $^{18}O_2$ show that in the formation of hydrogen peroxide or HO_2^- the ^{18}O–^{18}O bond remains intact. There is no exchange with $H_2^{16}O$ of the electrolyte. The four-electron route must, of course, go with cleavage of the oxygen molecular bond. The two routes (four- and two-electron reactions) are initiated by the manner in which the oxygen molecule adsorbs at the electrode surface. Side-on adsorption, in which bonds are formed with the surface through both oxygen atoms, leads to cleavage of the oxygen molecular bond; end-on adsorption, with only a single bond between oxygen and surface, leaves the molecular bond intact. Schemes which show the cooperation of the electrolyte have been proposed for the

Oxygen electrode reactions 221

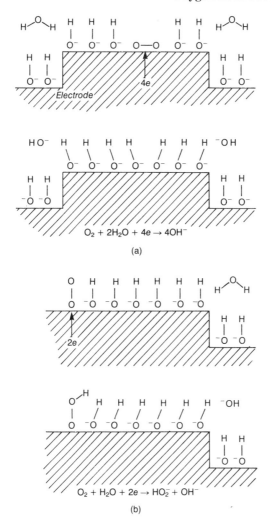

Figure 95. Possible scheme for the electrochemical reduction of oxygen in alkali by pseudo-splitting. (a) Side-on adsorption with the transfer of four electrons. (b) End-on adsorption with the transfer of two electrons

reduction of oxygen in alkali (*see* Figure 95). Catalysts that have been found to be particularly efficient in promoting the four-electron reaction are paramagnetic at the temperature of the experiments. It has been proposed that the surface electronic structure, having parallel spins, allows side-on adsorption.

Oxygen may also be reduced at elevated temperatures in a molten carbonate **fuel cell**.

222 Oxygen probe

Anodic reaction. The evolution of oxygen is the anodic reaction in the **electrolysis of water** and provides much of the **overpotential** of the cell. The overall reactions are

$$2H_2O \rightarrow O_2 + 4H^+ + 4e \quad \text{in acidic solution} \quad (e)$$

and

$$4OH^- \rightarrow O_2 + 2H_2O + 4e \quad \text{in alkaline solution} \quad (f)$$

The lack of reversibility of this system at metallic electrodes stems from the tendency of the electrode itself to be oxidized. Some correlation has been made between the oxidation potentials of metals and the potential above which oxygen is evolved at those metals. A simple scheme which involves a surface metal ion of oxidation state N (M^N) in alkali electrolyte is

$$M^N + OH^- \rightarrow M^{N+I}OH + e \quad (g)$$
$$M^{N+I} + OH^- \rightarrow M^{N+II}O + H_2O + e \quad (h)$$
$$2M^{N+II}O \rightarrow 2M^N + O_2 \quad (i)$$

The metal acts catalytically, and it may be seen that the potential at which oxygen is evolved will depend on the equilibrium potentials of reactions (g) and (h), rather than the overall potential derived from the free energy change of reactions (e) or (f). For example, on platinum PtO_2 (Pt^{IV}) is stable above 1.05 V at pH 14 (*see* **Pourbaix diagram**) and may be oxidized via Pt^V to the unstable PtO_3, which then chemically decomposes to PtO_2 and O_2. A reversible oxygen potential has been achieved at platinum by taking thin platinum foils saturated with oxygen. Changes occur in the platinum lattice as oxygen is alloyed with the metal and reversible behaviour is seen on the resulting surface.

The criterion of choosing a system with a potential of the metal–metal oxide half cell (or lower metal oxide–higher metal oxide half cell) which is near the thermodynamic potential of the oxygen electrode reaction has led to the development of new electrode materials such as the spinels $NiCo_2O_4$ and lithium-doped Co_3O_4.

Further reading
J. P. Hoare, *The Electrochemistry of Oxygen*, Wiley–Interscience, London (1968)

Oxygen probe. Two types of oxygen probe (galvanic and polarographic) are available; both are examples of a **gas-sensing membrane probe**, in which gaseous oxygen diffuses through a membrane so that the partial pressure of the gas is the same on both sides. In both types, a membrane (polyethylene or Teflon) holds the reference electrolyte in contact with the cathode or sensing element.

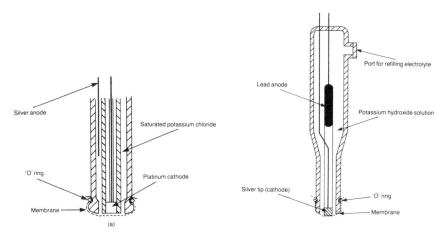

Figure 96. Schematic representation of oxygen probes: (a) Clark oxygen electrode (polarographic); (b) potentiometric

In the polarographic membrane probe (*see* Figure 96(a)), an external potential is applied across a non-polarizable reference electrode (silver) and a platinum microelectrode such that it is more negative than the reference electrode; the electrolyte is a saturated solution of potassium chloride. Oxygen is reduced at the cathode surface and a concentration gradient is set up leading to a diffusion of oxygen through the membrane towards the cathode. At a given voltage, a diffusion current, limited by the rate of diffusion of oxygen, is established. At the cathode

$$O_2 + 2H_2O + 4e \rightarrow 4OH^-$$

If the polarizing voltage is in the range 0.5–0.8 V, the diffusion current is proportional to the activity of oxygen in solution. This current can be amplified and fed to a chart recorder to give a trace of the change of concentration of oxygen with time.

In the galvanic probe (*see* Figure 96(b)), oxygen is reduced at a silver cathode and lead is consumed at the anode (the electrolyte is 1 mol dm^{-3} potassium hydroxide). At the anode

$$Pb + 4OH^- \rightarrow PbO_2^{2-} + 2H_2O + 2e$$

giving an overall cell reaction

$$O_2 + 2Pb + 4OH^- \rightarrow 2PbO_2^{2-} + 2H_2O$$

The current produced is passed through a standard resistance and the e.m.f. recorded on a digital voltmeter or a millivolt recorder. The oxygen concentration is proportional to the recorded potential.

Oxygen probe

When the aqueous solution flows past either type of probe, the dissolved oxygen diffuses across the membrane. Minimum flow conditions are critical since, if the flow rate is too low, the sample water around the electrode will be depleted of oxygen.

Both types of probe have a high temperature coefficient (about 6 percent per degree change); automatic temperature compensation is provided in modern instruments. The probes and measuring instrument must be calibrated before use in air-equilibrated water and in sodium sulphite solution. A correction for salinity must be applied when making determinations on sea water.

These probes are very versatile, finding use in monitoring oxygen concentrations in river and ocean water and in sewage, in following the respiration of isolated cell preparations and, in the miniaturized form, for monitoring oxygen in arteries.

P

Packed bed electrode. *See* **Fixed bed electrode.**

Passivity. A metal surface is said to be passive when, although exposed to conditions in which it is thermodynamically unstable, it remains unattacked indefinitely. Thus, iron is passivated by immersion in concentrated nitric acid; the initially vigorous attack of the metal ceases after a few seconds. Many other metals assume the passive state after comparable treatment.

Electrolytic passivation is illustrated in Figure 97, which shows the behaviour of an iron anode immersed in any suitable electrolyte when an increasingly positive potential is imposed on it. At first, the current increases in the usual way as metal ions go into solution, but at the passivation potential there is an abrupt change, and the current falls to a very low value which does not increase until some other anodic reaction takes over. The curve can be retraced, and the metal reactivated by holding it at a lower potential. Unless the potential change is very slow, the curve will show some hysteresis while the metal surface reverts to its original condition. Some passivation potentials, on the hydrogen scale, are: iron $+0.58$ V; nickel $+0.36$ V; chromium -0.22 V. Another commonly quoted potential is the Flade potential (*see* Figure 97), which is the potential corresponding to the point of inflexion in the current–voltage curve on the anodic side of the passivation potential.

Anodic protection, as a method of control of **corrosion**, consists of passivating the metal surface and maintaining it thus by imposing an external e.m.f. which supplies the small current passing under these conditions (a current very much less than that required for cathodic protection). Passivity has been attributed, since the time of Faraday, to an **oxide film**. In a few cases, the presence of this has been directly demonstrated, and the response of a passive electrode to pH changes is very much that of a metal–metal oxide electrode. On platinum, nickel and iron, the onset of passivity may only require a monolayer of oxide, or of adsorbed oxygen atoms, although a layer several nanometres thick may be produced in time. Evidence for these very thin layers, which are quite undetectable visually, comes from the small cathodic pulse of electricity that is sufficient to remove them, and from the phase and amplitude change of elliptically polarized light in the technique of **ellipsometry**. In other cases, for example lead, quite thick films are needed to preserve the passive state. Partial protection is obtained for many metals in a

226 Permittivity

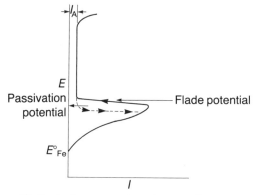

Figure 97. Passivation curve where I_A = anodic protection current

variety of electrolytes so long as conditions favour the precipitation of a film of insoluble hydroxide or salt (*see* **Anodizing**). These differences depend mainly on the nature of the surface layer. If its conduction is electronic, an anodic reaction such as oxygen evolution can proceed freely at its surface, leaving the thin oxide film unchanged. This is one extreme case; the other is that the film is porous to ions, in which case the attack of the metal will continue at a slow rate and the oxide film may grow to visible proportions.

Permittivity (ε; units: $F\ m^{-1} = A^2\ s^4\ kg^{-1}\ m^{-3}$). If two parallel conducting plates are electrified to surface charge densities $+\sigma$ and $-\sigma$, respectively, the field intensity between them, *in vacuo*, is given by

$$E_{vac} = \frac{4\pi\sigma}{\varepsilon_0} \tag{99}$$

where ε_0 is the permittivity of a vacuum. When the space between the plates is occupied by an insulating material, the field strength is reduced and is given by

$$E = \frac{4\pi\sigma}{\varepsilon} \tag{100}$$

where ε is always greater than ε_0. The permittivity of a material medium is related to the permittivity of a vacuum by

$$\varepsilon = \varepsilon_r \varepsilon_0$$

where ε_r is the relative permittivity, also known as the dielectric constant of the medium, this is a dimensionless quantity ($\varepsilon_0 = 8.854 \times 10^{-12}\ F\ m^{-1}$).

The permittivity of a solvent is a measure of its capacity for separating oppositely charged species. In a solvent with a high relative permittivity (e.g., water: $\varepsilon_r = 80.4$), a minimum amount of work is required to separate a positively charged ion from a negatively charged one. For a solvent with a low relative permittivity (e.g., acetic acid: $\varepsilon_r = 6.1$), a greater amount of energy is required to accomplish this process. The permittivity thus plays an important role in determining the strengths of solute acids or bases.

Per-salts. Anodic oxidation in aqueous solution often provides the best method of obtaining an element in its highest oxidation state. Ammonium and potassium persulphates (peroxydisulphate) are prepared by electrolysing concentrated solutions of the corresponding sulphates with cathodes usually of lead and smooth platinum anodes (at which the oxygen overpotential is high). The overall reaction is

$$2SO_4^{2-} \rightarrow S_2O_8^{2-} + 2e$$

but the system does not give a reversible potential, and several steps are probably involved.

Sodium perchlorate is made in a similar way by electrolysing concentrated sodium chlorate solutions between an iron cathode and a platinum anode

$$ClO_3^- + H_2O \rightarrow ClO_4^- + 2H^+ + 2e$$

Potassium permanganate is made by electrolysing potassium manganate between steel cathodes and nickel anodes

$$MnO_4^{2-} \rightarrow MnO_4^- + e$$

The cathode reaction gives hydrogen and potassium hydroxide, and the latter can be reutilized in the preparation of more manganate from the starting material, manganese dioxide.

pH. The negative logarithm (to base 10) of the hydrogen ion activity of the solution, that is

$$pH = -\log a_{H3O+}$$

Although the potential of hydrogen ion-indicating electrodes (hydrogen, glass, etc.) depends on a_{H3O+}, the impossibility of measuring an individual electrode potential precludes the determination of the pH value of a solution as defined. The modern definition of pH is an operational one, in which a suitable indicator electrode is combined with a reference electrode, and the e.m.f. of the cell so formed is determined. The e.m.f. values, E_X and E_S, of the cells

⊖ Pt,H₂(g) | solution X ⋮ KCl solution | Hg₂Cl₂, Hg ⊕
(1 atmos)

⊖ Pt,H₂(g) | solution S ⋮ KCl solution | Hg₂Cl₂, Hg ⊕
(1 atmos)

are measured at the same temperature; the concentration of the bridge potassium chloride solution should exceed 3.5 mol kg^{-1}. The e.m.f. values are

$$E_X = E_{cal} - E_{H+,H2} + E_{j,X} = E_{cal} - \frac{RT}{F} \ln a_{H3O+,X} + E_{j,X}$$

$$E_S = E_{cal} - E_{H+,H2} + E_{j,S} = E_{cal} - \frac{RT}{F} \ln a_{H3O+,S} + E_{j,S}$$

where E_{cal} is the electrode potential of the reference **calomel electrode**. Ignoring the two liquid junction potentials, $E_{j,X}$ and $E_{j,S}$, and subtracting these equations gives

$$E_X - E_S = \frac{RT}{F} \ln a_{H3O+,S} - \frac{RT}{F} \ln a_{H3O+,X}$$

$$= \frac{2.303 RT}{F} (pH_X - pH_S)$$

or

$$pH_X = pH_S + \frac{F(E_X - E_S)}{2.303 RT} \tag{101}$$

Now that the difference in pH of two solutions has been defined, the definition of pH_S can be completed by assigning, at each temperature, a value of pH_S to one or more chosen solutions designated as standards. The number assigned to pH_X may or may not have any theoretical significance. It has significance when $E_{j,X} = E_{j,S}$; this is believed to be true when $I < 0.1$ mol kg^{-1} and the pH is between 2 and 12. Under these conditions, the pH number has approximately the significance given by the formal definition.

Except in concentrated acid solutions, hydrogen ion activities are nearly equal to the hydrogen ion concentrations. Hence, for normal working conditions

$$pH = -\log c_{H3O+}$$

since $a_{H3O+} a_{OH-} = 10^{-14}$ mol^2 dm^{-6} in any dilute solution, at neutrality $a_{H3O+} = a_{OH-} = 10^{-7}$ mol dm^{-3}, and, hence, the pH value of a neutral solution is 7.0. Acid solutions have pH less than 7 and alkaline solutions have pH greater than 7. For a 0.01 mol dm^{-3} solution of a strong acid pH = 2; for a weak acid the pH depends on the concentration and the acid dissociation constant

$$\text{pH} = \tfrac{1}{2}pK_a - \tfrac{1}{2}\log(c/\text{mol dm}^{-3})$$

thus, the pH of a 0.1 mol dm^{-3} solution of acetic acid, of $pK_a = 4.756$, is 2.88.

pH values in non-aqueous solution, obtained from electrometric determinations (as for aqueous solutions), must be regarded simply as numerical values which may be of practical use for their reproducibility. Equations for the e.m.f. value of cells containing aqueous solutions are not strictly applicable to non-aqueous solutions because the high junction potential (e.g., at solvent/potassium chloride boundary) between solvents of different physical and chemical properties becomes less stable and less reproducible as the concentration of water decreases, and because of the difficulty of choice of a suitable reference electrode. The only satisfactory method is to consider all solvents as independent systems, without making any reference to aqueous systems. The hydrogen electrode in any solvent is always accepted as the standard electrode, while the reference electrode (usually a calomel electrode; a silver–silver chloride electrode is of limited use) contains the same non-aqueous solvent which is being used. The value of the 'pH' obtained has a purely conventional significance in that $E^{\ominus}_{H^+,H_2} = 0$ for all solvents, whereas in fact it almost certainly varies with the solvent. Glass, quinhydrone or antimony electrodes may be used in place of the hydrogen electrode.

Measurement of pH

1) Electrometric methods. Table 26 summarizes the characteristics, limitations and reproducibility of various hydrogen ion-indicating electrodes.

a) **Hydrogen electrode** and **calomel electrode**

$$\ominus \quad \text{Pt,H}_2(g) \mid \text{solution} \mid \text{KCl(aq)} \mid \text{Hg}_2\text{Cl}_2, \text{Hg} \quad \oplus$$
$$(101\,325 \text{ N m}^{-2})$$

$$E = E_{cal} + 0.0591 \text{ pH}$$

b) **Quinhydrone electrode** and **calomel electrode**

$$\ominus \quad \text{Hg,Hg}_2\text{Cl}_2 \mid \text{KCl(aq)} \mid \text{solution, saturated} \mid \text{Pt} \quad \oplus$$
$$\text{with quinhydrone}$$

$$E = E_{Q,QH2} - E_{cal} = E' - 0.0591 \text{ pH}$$

c) **Antimony–antimony oxide electrode** and **calomel electrode**

$$\ominus \quad \text{Sb, Sb}_2\text{O}_3 \mid \text{solution} \mid \text{KCl(aq)} \mid \text{Hg}_2\text{Cl}_2, \text{Hg} \quad \oplus$$

$$E = E' + 0.0591 \text{ pH}$$

d) **Glass electrode** and **calomel electrode**

$$\ominus \quad \text{Ag, AgCl} \mid \text{HCl} \mid \text{glass} \mid \text{solution} \mid \text{KCl(aq)} \mid \text{Hg}_2\text{Cl}_2, \text{Hg} \quad \oplus$$
$$(0.1 \text{ mol dm}^{-3})$$
$$\longleftarrow \text{glass electrode} \longrightarrow$$

Table 26

Characteristics of electrodes used for measurement of pH

Electrode	pH range	Interfering substances	Causes of error	Reproducibility/mV	Remarks
Hydrogen	0–14	Oxidizing and reducing agents, air, heavy metals	Incomplete saturation, O_2 in H_2, poisons	0.1	Follows theoretical equations, slow attainment of equilibrium, strong reducing action
Quinhydrone	1–8	Alkali, oxidizing and reducing agents, complexing agents, proteins	Poisons, salts	0.1	Follows theoretical equation, rapid attainment of equilibrium, solution contaminated
Antimony–antimony oxide	3–10	Strongly acid or alkaline conditions, H_2S, Cu^{2+}	Oxidizing agents organic compounds, salts, Cl_2, H_2S	5	Does not follow theoretical relationship, must be calibrated
Glass	0–12	Dehydrating agents, colloids, surface deposits	High concentration of alkali	0.1	Follows theoretical equation over certain pH range depending on type of glass, must be calibrated, equilibrium readily established, can be used in the presence of oxidizing and reducing agents

Because of the asymmetry potential of the glass electrode, which varies with time, the pH meter must be standardized with solutions of known pH.

2) **Indicators.** For an acid-base indicator,

$$HIn_A + H_2O \rightleftharpoons In_B^- + H_3O^+$$
acidic form　　　　　basic form
colour A　　　　　　colour B

$$pH = pK_{In} + \log \frac{c_{In,B-}}{c_{HIn,A}}$$

$$= pK_{In} + \log \frac{\text{intensity of colour B}}{\text{intensity of colour A}}$$

assuming that the concentrations of the two forms are proportional to the intensities of light transmitted by solutions of colours A and B. Thus, provided that pK_{In} is known, the pH of a solution can be obtained from a measurement of the ratio of intensities of the colours using a visual comparator or a photoelectric spectrophotometer.

In an alternative indicator method, the colour of the indicator in the solution of unknown pH is compared with that of standard buffer solutions containing the same concentration of indicator.

Photoelectrochemistry. The phenomenon resulting from the illumination of semiconductors in contact with liquid electrolytes. In photoelectrolysis, light is absorbed by the electrode with the promotion of an electron from valence to conduction band. The promoted electron may reduce a species in solution or the hole in the valence band may oxidize a species. A simplified diagram of the photoelectrolysis of water (which takes no account of band bending in the semiconductor) is given in Figure 98. The result of photoelectrolysis is therefore a new chemical substance. If the band gap of the semiconductor is less than the free energy change of the redox reaction, a voltage will need to be applied to effect the reaction.

In electrochemical photovoltaic cells, a similar situation prevails but with the output being electrical energy. In a photogalvanic cell, the light energy is absorbed by a dye (e.g., thionine) in solution. The excited dye reacts in a redox reaction (e.g., by providing the energy for $Fe^{2+} + H^+ \rightarrow Fe^{3+} + H$), and the products of this reaction are reoxidized and re-reduced at the anode and cathode, respectively. The net effect is transport of charge from one electrode to the other.

Certain photoelectrochemical reactions have been used to model photosynthesis. Synthetic chloroplasts have been prepared which consist of molecular pigments (e.g., methyl viologen + EDTA and ruthenium bisdipyridyl as sensitizer) associated with artificial membranes.

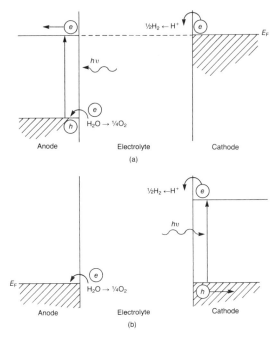

Figure 98. Energy diagram for two photoelectrolysis cells: (a) with photoanode (e.g., SrTiO$_3$) and metallic cathode; (b) with photocathode (e.g., WSe$_2$) and metallic anode

Further reading
Royal Society of Chemistry, Photoelectrochemistry, *Faraday Discussions of the Chemical Society*, no. 70 (1980)

Pitt's conductance equation. *See* **Conductance equations.**

Platinum and gold electrodes. Precious metal electrodes are used either as reference electrodes or for making electrical contact in oxidation–reduction systems (*see* **Redox electrode systems**). Platinum, the more commonly used metal, may be used in the shiny form (oxidation–reduction electrodes), the black form (hydrogen electrode and conductance cells to reduce polarization errors) or the grey form (conductance cells).

Platinum black electrodes, with an increased surface area, are prepared by electrolytically depositing a layer of platinum on the surface of a platinum electrode using a solution containing chloroplatinic acid and lead acetate. Platinum black can, in certain cases, have objectionable catalytic effects, and bright or greyed electrodes should be used. Grey electrodes are prepared by heating black electrodes to a dull red heat; this reduces the surface area. In very dilute solutions, blacked electrodes may cause a drift in the measured conductance readings because of the adsorption of electrolyte from solution. *See also* **Conductivity**.

Polarizable electrode. An ideal non-polarizable electrode is a system through which large currents can pass in either direction without diplacing the electrode potential from its equilibrium position. This is unattainable in practice, but electrode systems which have a high **exchange current density**, such as the $Ag^+|Ag$ electrode, can support large currents at quite a small **overpotential**.

The opposite extreme—the ideal 'polarizable electrode'—is one through which no current would pass whatever potential were imposed on the electrode. This condition is approached within certain potential limits by electrode–solution combinations in which no reversible charge transfer process is established.

Polarization. An electrode is said to be polarized whenever its potential departs from its equilibrium value. *See also* **Overpotential; Polarizable electrode.**

Polarography. A technique in which, under the influence of a potential applied to a dropping mercury electrode (DME), ions or molecules in the sample solution will tend to exchange electrons at the surface of the electrode; that is they will be oxidized if they lose electrons or reduced if they gain electrons. This results in a current consumption or production at the indicator electrode and when current is plotted against the applied potential a sigmoidal curve results.

The basic equipment and circuit for polarography is shown schematically in Figure 99. In commercial instruments, the process of increasing the voltage continuously and the measurement of the current flowing is fully automated to give a recording of the current–potential trace or 'polarogram' (*see* Figure 100). An indifferent electrolyte (e.g., potassium nitrate) is included in the test solution at a concentration of about 1 mol dm^{-3} to ensure that the solution has a high conductivity and that the indifferent electrolyte carries virtually all the current so that the electroactive species arrive at the DME by diffusion.

Dissolved oxygen, which is readily reduced at the DME giving two waves, must be displaced from the working solution by bubbling nitrogen (oxygen-free) through the solution. The broken line of Figure 100 is that which would be obtained in the absence of a reducible ion (i.e. with only the supporting electrolyte present); the final rise in the current is due to the discharge of the supporting electrolyte cation. The residual current is essentially a capacitance current due to the charging of the mercury drops as the double layer is formed. The shape of the polarogram (solid line) is often distorted by the so-called

234 Polarography

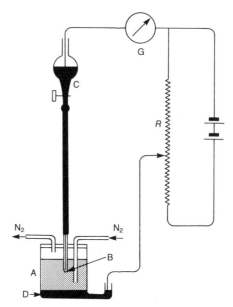

Figure 99. Principles and circuit of a simple polarograph: A, working solution containing depolarizer and supporting electrolyte; B, dropping mercury electrode (DME); C, mercury reservoir; D, mercury pool anode; G, galvanometer (used as a milliammeter); R, potentiometer for varying the applied potential between B and D

current maximum which interferes with the accurate evaluation of the diffusion current (I_d) and the half-wave potential ($E_{1/2}$). The exact cause of these maxima is not fully understood, but they may be eliminated by the addition of traces of high molecular weight substances (e.g., gelatin, Triton X-100).

The current plateau corresponds to the condition that the metal ions are reduced as fast as they reach the electrode (i.e. complete concentration

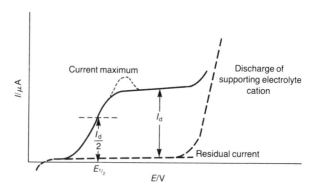

Figure 100. Typical polarogram

polarization). An electrode is polarized when its potential may be altered without appreciable change in the current flowing, the condition obtaining at the foot and plateau of the polarogram. On the rising part of the curve, the electrode becomes depolarized by the reduction process.

The magnitude of the diffusion-controlled current depends on the concentration of the electroactive species in solution, and this forms the basis of the use of polarography as an analytical technique, expressed in the Ilkovic equation

$$I_d = 607\, n\, D^{1/2}\, m^{2/3}\, t^{1/6}\, c \tag{102}$$

where I_d μA is the time-average diffusion current during the lifetime of the drop, n the number of electrons exchanged in the electrode reaction, D cm^{2}s^{-1} the diffusion coefficient of the reactive species, m mg s^{-1} the rate of mass flow of the mercury, t s the drop time and c mmol dm^{-3} the concentration of the depolarizer. The capillary factor $m^{2/3}\, t^{1/6}$ describes the influence of the dropping electrode characteristics on the diffusion current. The most temperature-sensitive factor in the Ilkovic equation is the diffusion coefficient which changes by about 2.5 percent per degree; as a consequence, good temperature control is essential for accurate polarographic analysis.

The half-wave potential ($E_{1/2}$) at the mid-point of the curve ($I = I_d/2$) is characteristic of a given depolarizer for defined experimental conditions and may permit qualitative identification of the electroactive species. If the reduction process occurs reversibly, the relationship between the current and the potential on the rising part of the curve is explained in the Heyrovsky–Ilkovic equation

$$E = E_{1/2} + \frac{RT}{nF} \ln \frac{I_d - I}{I} \tag{103}$$

Eq. (103) is similar to the **Nernst equation** in which $E^\ominus = E_{1/2}$; the plot of ln $(I_d - I)/I$ against E may be used to determine n and give an accurate value of the half-wave potential.

For irreversible reductions, the slope of the wave is less than the theoretical value and a **transfer coefficient** α is introduced

$$E = E_{1/2} + \frac{RT}{\alpha nF} \ln \frac{I_d - I}{I}$$

for such systems $E_{1/2} > E^\ominus$ and the electron transfer process is rate-determining at the start of the wave, towards the later part of the curve the diffusion process takes over as rate-determining.

Both reversible and irreversible diffusion-controlled waves can be used for analytical purposes; under standard conditions of test (i.e. constant D, m and t; Eq. (102)), the limiting current is diffusion-controlled and the linear calibration curve of current against concentration can be used to determine an

236 Polarography

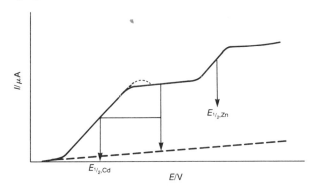

Figure 101. Polarogram for a mixture of zinc and cadmium ions

unknown concentration. Alternatively, an unknown concentration can be obtained, by simple proportion, from the increased current resulting from the addition of a known concentration of an internal standard to the test solution.

In a mixture of electroactive species, each will behave independently and a series of waves will result (*see* Figure 101), the height of each wave being directly proportional to the concentration of the appropriate species. In the presence of a complexing agent, the value of the half-wave potential varies with the concentration of the complexing agent; this can be used to determine the formula and the formation constant of the complex.

Normal pulse polarography, which gives a marked improvement in sensitivity over the standard dc polarography, still gives a sigmoid-shaped polarogram. Differential pulse polarography, in which small amplitude pulses (10–100 mV) each of about 60 ms duration are superimposed on a conventional dc voltage and applied to the DME near the end of the lifetime of the drop, gives a further improvement in sensitivity (10^{-8} mol dm^{-3} compared with 10^{-5} mol dm^{-3} for normal dc techniques). In this method, the current is measured 17 ms before the pulse and at the end of the pulse, ΔI is recorded as a function of the applied potential.

Applications
1) Inorganic polarographic analysis. Polarographic methods are widely used in the analysis of inorganic substances. Most cations are reduced at the DME to form a metal amalgam or ions of a lower valence type. The method is also applicable to such anions as bromate, iodate, dichromate, nitrite, tungstate, vanadate and molybdate (the limits of detection are about 10^{-7}–10^{-8} mol dm^{-3}) and has been used to determine gaseous oxygen in gas mixtures, biological fluids and water.
2) Organic polarographic analysis. Several common functional groups are oxidized or reduced at the DME, and compounds containing these groups can

be analysed by polarography. In general, the reactions of organic compounds at DME are slower and more complex than those for the simple inorganic ions. Although theoretical interpretation of data is more difficult, nevertheless organic polarography has proved useful for the determination of structures and for the qualitative identification and quantitative analysis of components in mixtures.

Since protons are commonly involved in organic oxidation–reduction processes, good buffering of the test solution is necessary for the production of accurate and reproducible half-wave potentials and current values. On account of solubility problems, water is not always a suitable solvent and aqueous mixtures of glycols, dioxane and alcohols are often used. Anhydrous media, such as glacial acetic acid, formamide and ethylene glycol have been investigated; supporting electrolytes are often lithium salts or tetraalkyl-ammonium salts. Organic compounds containing any of the following functional groups produce one or more polarographic waves: $>C=O$ (aldehydes, ketones, quinones); nitro; nitroso; halogen; conjugated $C=C$.

Polarographic methods are now widely used in the analysis of naturally occurring mycotoxins (e.g., fungal metabolites), compounds used to improve food yields (e.g., insecticides and fungicides) and environmental pollutants (e.g., N-nitrosoamines, detergents, azo dyes). See also **Stripping voltammetry**.

Further reading
G. Dryhurst, K. M. Kadish, F. Scheller and R. Renneberg, *Biological Electrochemistry*, vol. I, Academic Press, New York (1982)
B. Fleet and R. D. Jee, in *Electrochemistry*, vol. 3, *Electroanalytical Chemistry: Voltammetry*, Specialist Periodical Reports, Chemical Society, London, p. 210 (1973)
W. Franklin-Smyth (ed.), *Polarography of Molecules of Biological Significance*, Academic Press, London (1979)
W. Franklin-Smyth (ed.), *Electroanalysis in Hygiene, Environmental, Clinical and Pharmaceutical Chemistry*, Elsevier, Amsterdam (1980)
J. Heyrovsky and P. Zuman, *Practical Polarography*, Academic Press, London (1968)
J. Volke, Polarographic and voltammetric methods in pharmaceutical chemistry and pharmacology, *Bioelectrochem. Bioenerget.*, **10**, 7 (1983)
P. Zuman, *Organic Polarographic Analysis*, Pergamon Press, Oxford (1964)

Porous electrode. Electrode reactions occur in two dimensions only and thus have an inherent, spatial disadvantage compared with reactions in bulk solution. The available surface area of an electrode may be increased by using high surface area catalysts in a porous electrode. Thus some three dimension-

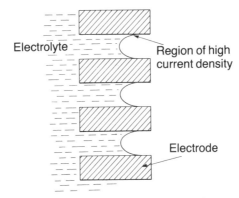

Figure 102. Magnified cross-section of an idealized porous electrode

ality is imparted to the electrode. The increased maximum current available from a porous electrode arises not from the increased internal surface area, but from reductions in the thickness of the diffusion layer (*see* **Diffusion-limited current**). It may be shown that the current is concentrated at the mouth of the pore (*see* Figure 102), but the meniscus gives a thin diffusion layer and hence greater limiting current density. An ideal porous electrode would attempt to coincide the position of the meniscus with a region of active catalyst material. *See also* **Teflon-bonded electrode**.

Potential (V). At any point the potential is measured by the work necessary to bring a unit positive charge from an infinite distance. The potential at a point due to a charge q at a distance r in a medium of relative permittivity ε_r is

$$V = \frac{q}{4\pi\varepsilon_0\varepsilon_r r}$$

See also **Interfacial potential; Outer potential; Surface potential.**

Potential distribution. An advantage of electrochemical syntheses is the ability to select certain reactions by virtue of the potential at which they occur. This is only feasible when the potential is uniform over the whole area of the electrode. Nonuniform potential distribution arises from (1) nonuniform **current distribution** resulting from differences in the concentrations of reactants across the electrode, (2) electrode resistance leading to an ohmic potential loss across the electrode away from the current collector and (3) potential drop through the electrolyte. This is most evident for electrolyte trapped in pores (*see* **Porous electrode**). The potential distribution in the cell depends largely on the electrode geometry, and is only uniform if the arrangement of the electrodes is highly symmetrical: for example, for parallel-plate electrodes or concentric cylinder electrodes.

The potential distribution is given by the solution of the Laplace equation

$$\nabla^2 \phi = 0$$

for the solution and electrode.

Potential of zero charge. The potential at which the charge on a polarizable electrode is zero is known as the potential of zero charge (ϕ_{pzc}). ϕ_{pzc} is determined from electrocapillarity measurements (*see* **Electrocapillary phenomena**), and is the potential at which the slope $\partial \gamma / \partial E$ is zero. At the potential of zero charge, the absolute potential difference between the electrode and electrolyte is equal to the **surface potential** difference ($\Delta \chi$). The current–voltage characteristics are most sensitive to the state of the double layer (*see* **Electrical double layer**) near the potential of zero charge. The source of the correlation between work function and electrocatalytic activity (*see* **Electrocatalysis**) is the dependence of the potential of zero charge on the work function of the metal (*see* Figure 103). See also **Isopotential point**.

Potentiometric titrations. See **Electrometric titrations**.

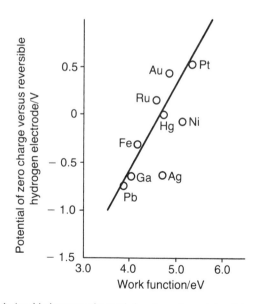

Figure 103. Relationship between the work function of a metal and its potential of zero charge

240 Potentiostat

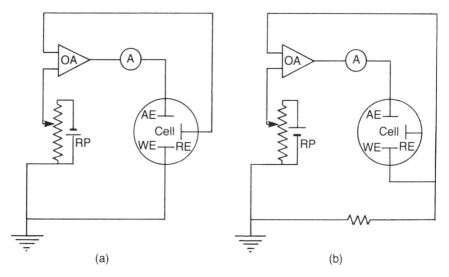

Figure 104. Circuit diagram for a potentiostat operating in (a) potentiostatic and (b) galvanostatic mode. OA, operational amplifier; A ammeter; RP, reference potential; AE, auxillary electrode; WE, working electrode; RE, reference electrode

Potentiostat. A device which allows current to flow between a working and secondary electrode (*see* **Three-electrode system**) while maintaining a constant voltage between the working electrode and reference electrode (*see* Figure 104(a)). A potentiostat is essentially a high-gain differential amplifier which measures the potential between the working and reference electrodes, and compares it to a set reference voltage provided by a potentiometer. Current is passed to bring this difference to zero.

A potentiostat may be used in galvanostatic mode by including a known resistance between working and reference electrodes (*see* Figure 104(b)). A constant voltage is developed across the resistor by the potentiostat and by Ohm's law a constant current flows in the circuit (*see* **Galvanostat**).

Pourbaix diagram. A graph of the limits of stability of different solid phases and oxidation states of a metal as a function of potential and pH. The example of nickel is given in Figure 105. In calculating the pH at which a solid phase may dissolve, for example

$$NiO + 2H^+ \rightarrow Ni^{2+} + H_2O$$

a value of 10^{-6} mol dm^{-3} for the concentration of the soluble ion is taken to correspond to a reasonable amount of dissolution. The equation given above is independent of potential (no electrons are involved in the stoichiometry) and this is represented on the diagram by a vertical line. A horizontal line

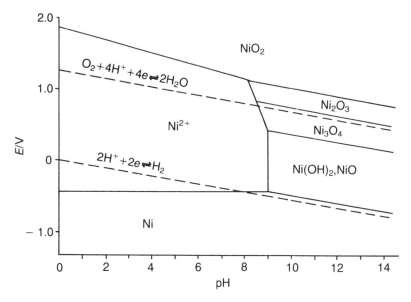

Figure 105. Pourbaix diagram of nickel

indicates no pH dependence, for example, for the reaction $Ni^{2+} + 2e \rightarrow Ni$. Superimposed on the diagram are the equilibrium potentials of the hydrogen electrode and the oxygen electrode. The relationship of a phase at given pH and potential to these lines shows that if an electron path is available **corrosion** may occur with evolution of hydrogen or reduction of oxygen as the electronation reaction. With reference to nickel, it is seen that the line dividing Ni and Ni^{2+} is below the hydrogen line for pH values up to pH 9. Thus nickel should corrode in these solutions by the reactions

$$Ni \rightarrow Ni^{2+} + 2e$$

and

$$2H^+ + 2e \rightarrow H_2$$

or, as the Ni, Ni^{2+} line is also below the oxygen line,

$$O_2 + 4H^+ + 4e \rightarrow 2H_2O$$

Pourbaix diagrams only give information concerning the thermodynamic stability of the phases. Nickel passivates at +0.36 V versus standard hydrogen electrode, and so remains unaffected even in regions of pH and potential in which, according to the Pourbaix diagram, it is unstable (*see* **Passivity**).

Power density. *See* **Electrochemical storage.**

Primary battery. *See* **Electrochemical storage; Leclanché cell.**

Proton jump transport. *See* **Molar ionic conductivity.**

Pseudocapacitance. Consider an electrochemical reaction in which an adsorbed intermediate I_{ads} is formed

$$ox + e \rightarrow I_{ads}$$
$$I_{ads} \rightarrow red$$

A constant current switched on at time $t = 0$ leads to the charging of the double layer and the formation of a layer of I_{ads}. The double-layer differential capacitance ($\tilde{C}_D$) is defined in terms of the charge passed (q_D) and change in electrode potential ($\Delta\phi$)

$$\tilde{C}_D = \frac{dq_D}{d\Delta\phi} = \frac{I_D}{d\Delta\phi/dt}$$

where I_D is the double-layer charging current. The charge dq_I, which forms the adsorbed layer, may be related to the fractional coverage $d\theta$

$$dq_I = I_I dt = q_m d\theta$$

where q_m is the charge required to form a monolayer of I_{ads}. Therefore

$$\frac{dq_I}{d\Delta\phi} = \frac{I_I}{d\Delta\phi/dt} = q_m \frac{d\theta}{d\Delta\phi}$$

The terms $dq_I/d\Delta\phi$ and $I_I/(d\Delta\phi/dt)$ have the units of capacitance and are called the adsorption pseudocapacitance or, simply, the pseudocapacitance ($\tilde{C}_I$). The total current is therefore

$$I = (\tilde{C}_D + \tilde{C}_I)\frac{d\Delta\phi}{dt} + I_{CT}$$

where I_{CT} is the current due to charge transfer, that is the electrochemical reaction. Any measurement of capacitance in a reaction involving an adsorbed intermediate will thus include a pseudocapacitance. *See also* **Capacitance.**

Q

Quinhydrone electrode. A reversible **redox electrode system** which acts as a hydrogen ion-indicating electrode. It consists of a shiny platinum electrode dipping in the test solution which is saturated with quinhydrone (i.e. an equimolar mixture of quinone (Q) and hydroquinone (QH_2) where $Q = C_6H_4O_2$. The electrode reaction is

$$Q + 2H^+ + 2e \rightleftharpoons QH_2$$

and its **electrode potential** is given by

$$E_{Q,QH2} = E^\ominus_{Q,QH2} + \frac{RT}{2F} \ln \frac{a_Q a_{H+}^2}{a_{QH2}}$$

The ratio a_Q/a_{QH2} is constant and equal to unity provided that the equilibrium is not disturbed by the presence of other oxidation–reduction systems, hence

$$E_{Q,QH2} = E'_{Q,QH2} + \frac{RT}{F} \ln a_{H+}$$

The quinhydrone electrode therefore acts as a hydrogen ion-indicating electrode at pH values up to 8; above this pH, the ratio a_Q/a_{QH2} no longer remains constant and atmospheric oxidation occurs. The electrode has a salt error and cannot be used in the presence of oxidizing or reducing agents, amino compounds, ammonia and ammonium salts. See Ives and Janz (1961).

R

Redox electrode systems. Although all half cells are really oxidation–reduction electrodes (*see* **Half cell**), the term 'redox' is restricted to an inert platinum electrode in contact with the oxidized and reduced forms of an electrode couple in the cell solution: for example, $Fe^{3+}, Fe^{2+} | Pt$. The e.m.f. of the cell

$$\ominus \quad Pt, H_2(g) \quad | \quad H^+ \quad | \quad ox, red \, | \, Pt \quad \oplus$$
$$(101\,325 \text{ N m}^{-2}) \quad (a_{H^+} = 1)$$

(in which the **liquid junction potential** has been eliminated and hence the redox electrode potential) is given by the **Nernst equation**

$$E = E^{\ominus}_{ox,red} + \frac{RT}{nF} \ln \frac{a_{ox}}{a_{red}} \qquad (98)$$

The standard electrode potential is the potential when the activity of the oxidized form is equal to the activity of the reduced form. If the individual activities (a_{ox} and a_{red}) are not available by independent methods, the Nernst equation may be written as

$$E = E' + \frac{RT}{nF} \ln \frac{c_{ox}}{c_{red}} \qquad (104)$$

where E', the formal electrode potential, given by

$$E' = E^{\ominus}_{ox,red} + \frac{RT}{nF} \ln \frac{\gamma_{ox}}{\gamma_{red}} \qquad (105)$$

depends on the **ionic strength** of the solution. If E is measured at different values of the ratio c_{ox}/c_{red}, the graph of E against $\ln c_{ox}/c_{red}$ is linear and of slope RT/nF; the intercept when $\ln c_{ox}/c_{red}$ is zero gives the value of E' at the particular ionic strength of the solution. If α is the proportion of the oxidized form, Eq. (104) becomes

$$E = E' + \frac{RT}{nF} \ln \frac{\alpha}{1-\alpha} \qquad (106)$$

The graph of E against α depends on the value of n (*see* Figure 106); a point of inflexion occurs at $\alpha = 0.5$, the tangent at this point is $4RT/nF$.

From a knowledge of $E^{\ominus}_{ox,red}$, the potential of any mixture of oxidized and reduced forms can be calculated.

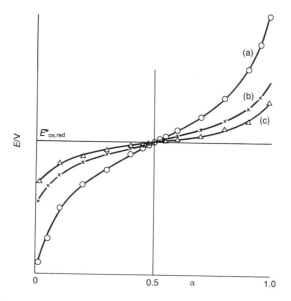

Figure 106. Variations of electrode potential with α (fraction oxidized): (a) $n = 1$; (b) $n = 2$; (c) $n = 3$

The redox potential is the potential adopted by the platinum electrode due to the equilibrium

$$\text{ox} + n\text{e} \rightleftharpoons \text{red}$$

If the equilibrium lies to the left (right), the inert electrode adopts a negative (positive) potential and the redox system is a good reducing (oxidizing) agent. The more positive the redox potential of a system, the better oxidizing agent is that system. Thus a system with a more positive potential will oxidize one with a less positive system (electrons are transported in the direction of increasing electrode potential).

Determination of standard redox potentials.
1) Using a simple cell with a reference electrode, as for a normal **electrode potential**. For example, the cell

$$\ominus \text{ Ag,AgCl} \,|\, \text{Cl}^-, \text{Fe}^{2+}, \text{Fe}^{3+} \,|\, \text{Pt } \oplus$$

has the cell reaction

$$\text{Fe}^{3+} + \text{Ag(s)} + \text{Cl}^- \rightleftharpoons \text{Fe}^{2+} + \text{AgCl(s)}$$

for which the e.m.f. is given by

$$E - \frac{RT}{F} \ln \frac{c_{\text{Fe}3+} c_{\text{Cl}-}}{c_{\text{Fe}2+}} = E^\ominus_{\text{cell}} + \frac{RT}{F} \ln \frac{\gamma_{\text{Fe}3+} \gamma_{\text{Cl}-}}{\gamma_{\text{Fe}2+}} \qquad (107)$$

246 Redox electrode systems

From a knowledge of the e.m.f. of the cell at different concentrations of the oxidized and reduced forms, the graph of the right-hand side of Eq. (107) against the square root of the ionic strength is linear and of intercept

$$E^\ominus_{cell} = E^\ominus_{Fe^{3+},Fe^{2+}} \quad E^\ominus_{AgCl,Ag,Cl^-}$$

This is not a good method, since exact knowledge of the concentrations of the oxidized and reduced forms is difficult to obtain.

2) From equilibrium constant measurements. This is probably the most accurate method. The cell reaction of the hypothetical cell

$$\text{Pt} \mid \text{Fe}^{2+}, \text{Fe}^{3+} \mid \text{Ag}^+ \mid \text{Ag}$$

is

$$\text{Fe}^{2+} + \text{Ag}^+ \rightleftharpoons \text{Fe}^{3+} + \text{Ag(s)}$$

and the standard e.m.f. is given by

$$E^\ominus_{cell} = \frac{RT}{F} \ln K_{therm} = \frac{RT}{F} \ln \left(\frac{a_{Fe^{3+}}}{a_{Fe^{2+}} a_{Ag^+}} \right)_e \quad (108)$$

$$= E^\ominus_{Ag^+,Ag} - E^\ominus_{Fe^{3+},Fe^{2+}}$$

The equilibrium constant K_c is calculated from measurements of the equilibrium concentrations of iron(II), iron(III) and silver ions when a solution of iron (III) perchlorate containing excess perchloric acid (to prevent hydrolysis) is shaken with finely divided metallic silver at different ionic strengths.

$$K_{therm} = K_c \frac{\gamma_{Fe^{3+}}}{\gamma_{Fe^{2+}} \gamma_{Ag^+}}$$

since

$$\log \gamma_i = -Az_i^2 I^{1/2} + CI$$

it follows that

$$\log K_{therm} = \log K_c - 4AI^{1/2} + CI$$

Thus from measured values of K_c, $E^\ominus_{cell}$ and hence $E^\ominus_{ox,red}$ for the redox system can be calculated.

3) From potentiometric titrations (*see* **Electrometric titrations**). This method, mainly used for systems involving organic and biological compounds, only allows the determination of formal electrode potentials since activity coefficients are ignored. It is usually impossible to make the correction for these to yield a true standard electrode potential on account of the high ionic strengths used.

The pure oxidized (reduced) form of the substance, dissolved in a buffer solution of known pH and ionic strength, is titrated against a solution of a

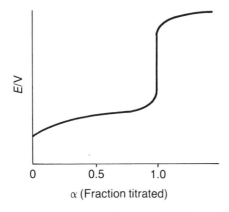

Figure 107. Variation of e.m.f. of cell during titration

reducing (oxidizing) agent in the complete absence of air. After each addition of titrant, the e.m.f. of the cell, comprising a platinum and a reference electrode, is measured and plotted as a titration curve (*see* Figure 107). For the cell

$$\ominus \; Hg,Hg_2Cl_2 \; \Big| \; \text{saturated } KCl(aq) \; \Big| \; red,ox \; \Big| \; Pt \; \oplus$$

$$E_{cell} = E_{ox,red} - E_{cal} = E' - E_{cal} + \frac{RT}{nF} \ln \frac{c_{ox}}{c_{red}}$$

$$= E' - E_{cal} + \frac{RT}{nF} \ln \frac{\alpha}{1 - \alpha}$$

(from Eq. (104) and Eq. (106)). The graph of E_{cell} against $\ln [\alpha/(1 - \alpha)]$ is linear (*see* Figure 108), of intercept $E' - E_{cal}$ and of slope RT/nF enabling the determination of the number of electrons in the oxidation process.

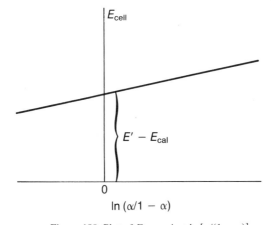

Figure 108. Plot of E_{cell} against $\ln [\alpha/(1 - \alpha)]$

Oxidation–reduction systems play an important role in biological processes involving electron transport. *See also* **Biological oxidation–reduction systems; Electron transport chain;** Appendix, Tables 5 and 6. See James and Prichard (1977).

Reference electrode. A reversible, non-polarizable half cell at which the potential remains in equilibrium, against which the potential of a test half cell may be measured or controlled. Common reference electrodes are given elsewhere (*see* Appendix, Table 11). *See also* **Electrometric titrations; Electrode potential; Fused salts; Organic electrochemistry; Three-electrode system.** See Ives and Janz (1961).

Relative permittivity. *See* **Permittivity;** Appendix, Table 2.

Resistance overpotential. *See* **Overpotential.**

Rest potential. *See* **Open-circuit voltage.**

Reversible electrode potential. *See* **Electrode potential.**

Reversible galvanic cell. Any arrangement, consisting of two electrodes and an electrolyte, capable of undergoing spontaneous chemical reaction to produce an electric current when the electrodes are joined externally, is called a galvanic cell. Cells of this type may be divided into two categories, depending on whether a chemical reaction occurs even before there is a flow of current (irreversible cells) or whether there is no reaction until the electrodes are connected externally and there is a current flow (reversible cells).

An example of an irreversible cell is

$$Zn \mid \text{dilute } H_2SO_4 \mid Pt$$

in which the zinc reacts spontaneously with the acid even before the electrodes are connected externally. Cells of this sort are always irreversible in the thermodynamic sense and are not of much theoretical interest.

The reversibility of galvanic cells can be tested by connecting the cell under consideration to an external source of e.m.f. which is adjusted so as to balance the e.m.f. of the cell exactly. Under these conditions no current flows, and there is no chemical change in the cell. If the external e.m.f. is decreased by an infinitesimal amount, current will flow from the cell and a chemical change, proportional to the quantity of electricity passed, should occur. On the other hand, if the opposing e.m.f. is slightly greater than that of the cell, current will flow in the opposite direction and the cell reaction will be reversed. Both electrode reactions of a reversible cell must be reversible, that is both electrodes must be reversible.

Such electrochemical cells will only behave reversibly when the current flowing is infinitesimally small and the system is virtually always in equilibrium. If large currents flow, concentration gradients are set up and the cell can no longer be regarded as in a state of equilibrium. It is for this reason that high resistance digital voltameters are used to measure the e.m.f. of such cells.

The e.m.f. of a cell varies with the concentration and, hence, the **activity** of the electrolyte solution, the gas pressure for a **gas electrode** and the amalgam concentration for an **amalgam electrode**.

For the simple galvanic cell consisting of the **hydrogen electrode** in combination with the **silver–silver chloride electrode**

$$\ominus \text{ Pt,H}_2(\text{g}) \mid \text{H}^+\text{Cl}^- \mid \text{AgCl,Ag } \oplus$$
$$(p \text{ atmos})$$

The electrode and cell reactions are

$$\tfrac{1}{2}\text{H}_2(\text{g}) \rightarrow \text{H}^+ + e$$
$$\text{AgCl(s)} + e \rightarrow \text{Ag(s)} + \text{Cl}^-$$

$$\overline{\tfrac{1}{2}\text{H}_2(\text{g}) + \text{AgCl(s)} \rightarrow \text{Ag(s)} + \underset{(m)}{\text{H}^+ + \text{Cl}^-}}$$

For this cell, as written, the e.m.f. is positive and, hence, the forward reaction is spontaneous. For this cell reaction, the free energy change (ΔG) is given by the van't Hoff isotherm

$$\Delta G = \Delta G^\ominus + RT \ln \frac{a_{\text{Ag(s)}} a_{\text{H}^+} a_{\text{Cl}^-}}{a_{\text{H2(g)}}^{1/2} a_{\text{AgCl(s)}}} \tag{109}$$

but

$$\Delta G = -nFE \tag{110}$$

Combining Eq. (109) and Eq. (110), and replacing $a_{\text{H2(g)}}$ by p_{H2}, remembering that silver and silver chloride are in their standard states, gives

$$E = E^\ominus - \frac{RT}{F} \ln a_{\text{H}^+} a_{\text{Cl}^-} + \frac{RT}{F} \ln p_{\text{H2}}^{1/2}$$

$$= E^\ominus - \frac{2RT}{F} \ln a_\pm + \frac{RT}{F} \ln p_{\text{H2}}^{1/2} \tag{111}$$

Alternatively, the e.m.f. of the cell may be obtained from the algebraic sum of the two electrode potentials

$$E = E_{\text{AgCl,Ag,Cl}^-} - E_{\text{H}^+,\text{H2}}$$

$$= E^\ominus_{\text{AgCl,Ag,Cl}^-} - \frac{RT}{F} \ln a_{\text{Cl}^-} - E^\ominus_{\text{H}^+,\text{H2}} - \frac{RT}{F} \ln \frac{a_{\text{H}^+}}{p_{\text{H2}}^{1/2}}$$

$$= E^{\ominus}_{\text{cell}} - \frac{2RT}{F} \ln a_{\pm} + \frac{RT}{F} \ln p_{\text{H}_2}^{1/2} \tag{111}$$

When the pressure of hydrogen is kept constant at 1 atmos, Eq. (111) can be written

$$E = E^{\ominus}_{\text{cell}} - \frac{2RT}{F} \ln m_{\pm} - \frac{2RT}{F} \ln \gamma_{\pm} \tag{112}$$

Thus the plot of $E + 2RT/F \ln m_{\pm}$ against $I^{1/2}$ is linear and of intercept $E^{\ominus}_{\text{cell}} = E^{\ominus}_{\text{AgCl,Ag,Cl}^-}$.

The e.m.f. of cells consisting of a hydrogen and a silver, silver halide in different solvents

$$\text{Pt,H}_2(g) \,|\, \text{HX—S} \,|\, \text{AgX,Ag}$$

depends on the solvent S. From such studies, it is possible to determine the standard electrode potentials in the different solvents (*see* Table 27). In general, the standard electrode potentials in all the solvents studied become more negative in the order chloride, bromide and iodide.

Table 27
Standard electrode potentials of $X^- \,|\, \text{AgX,Ag}$ at 298 K in a range of solvents

Solvent	$E^{\ominus}/V$		
	$Cl^- \,|\, \text{AgCl,Ag}$	$Br^- \,|\, \text{AgBr,Ag}$	$I^- \,|\, \text{AgI,Ag}$
Water	+0.2224	+0.073	−0.151
Methanol	−0.0101	−0.1357	−0.3176
Ethanol	−0.08138	−0.1816	−0.2530
Dimethyl sulphoxide	+0.1288		
Formamide	+0.2002		
Ethylene glycol	+0.0235	−0.098	−0.2928
Glacial acetic acid	−0.6208		

Galvanic cells of the type

$$\text{M(Hg)} \,|\, \text{MX—S} \,|\, \text{AgX,Ag}$$

where M is an alkali metal, have been studied in a range of non-aqueous solvents, leading to values for the standard electrode potentials of alkali metals (*see* Appendix, Table 4).

Rotating-disc electrode (RDE). The current at an electrode is determined by the following factors: (1) transport of species to and from the electrode; (2) kinetics of the electron transfer process; (3) kinetics of any chemical (homogeneous or heterogeneous) reactions. To obtain reproducible currents,

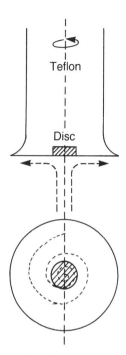

Figure 109. Flow lines around a rotating-disc electrode

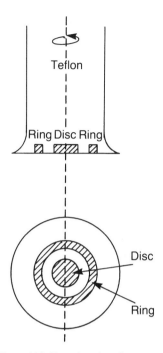

Figure 110. Rotating-ring disc electrode

the transport effects must be known or controlled. At a stationary planar electrode, the transport near the electrode may only be accurately known and controlled for short times. In the technique of **polarography**, the transport problem is resolved by having a continuously growing mercury drop around which the diffusion characteristics of the solution are well known. The solution of the diffusion equations is a cyclic one in time as the surface area of the drop grows and falls. The use of the RDE leads to well-known, stationary solutions for the movement of electrolyte near the electrode surface. The RDE consists of a disc of electrode material set in an inert material such as Teflon which is rotated about its axis at a known speed. Figure 109 shows the way the electrolyte moves near a RDE. The thickness of the diffusion layer (z_D) is

$$z_D = 0.643 \, W^{-1/2} \, \nu^{1/6} \, D^{1/3}$$

where W is the rotation speed in Hz, ν the kinematic viscosity (cm² s⁻¹) and D is the diffusion coefficient (cm² s⁻¹). The limiting current density (*see* **Diffusion-limited current**) is given by the Levich equation

$$i_L = 1.554 \, W^{1/2} \, \nu^{-1/6} \, D^{2/3} \, c_{bulk}$$

where c_{bulk} is the concentration of the reactant in the bulk of the solution.

Advantages of the RDE over a dropping mercury electrode are: (1) any metal may be used as the electrode material, thus allowing the use of high anodic potentials; (2) a genuine steady-state current is produced; (3) the theory is more simple and accurate; (4) the rotation speed is a more easily controlled variable than the drop time; (5) faster kinetics may be studied.

A most useful extension of the RDE is to add an annulus electrode which is insulated from the disc (*see* Figure 110). The annulus of the rotating-ring disc electrode (RRDE) may be held at any desired potential such that, for example, the current at the ring reflects the oxidation or reduction of some intermediate of the reaction at the disc which has been transported to it. Used in this way, a RRDE has detected copper (I) ions in the reduction of copper (II) ions to copper and hydrogen peroxide in the reduction of oxygen. The RRDE may also be used to investigate the kinetics of fast homogeneous reactions if one of the reactants can be produced electrochemically at the disc, and the reactants or products monitored at the ring. The reaction between bromine, generated from sodium bromide, and allyl alcohol was investigated by this method. A further refinement occurs if the ring is split allowing two separate measurements to be made at the ring.

Further reading

W. J. Albery and M. L. Hitchman, *Ring Disc Electrodes*, Oxford University Press, Oxford (1971)

S

Sacrificial protection. *See* **Corrosion.**

Salt bridge. A method of connecting a **reference electrode** with the test solution or two solutions such that the **liquid junction potential** is reduced as far as possible. A salt bridge consists of an inverted U-tube plugged at the ends either with sintered discs of fine porosity or with tightly rolled spirals of filter paper. The tube is filled with the required bridge solution and the bridge stored with the legs dipping in vessels containing the same solution. Bridges of this type have the advantage over agar salt bridges that they last indefinitely and do not suffer from syneresis or shrinkage of the gel. Where contamination of the test solution with the bridge solution must be kept to a minimum, an agar bridge must be used.

Depending on the nature of the solutions to be connected, the electrolytes in salt bridges may be potassium chloride or nitrate or ammonium nitrate.

Sea water battery. A primary cell which is stored dry and is activated by immersion in sea water. Sea water batteries have applications in marine safety lights, buoys, flashing beacons, torpedoes, mines, etc. Magnesium and zinc are common anode materials with silver chloride or copper (I) chloride as cathodes. The cell reactions are thus

$$\text{cathode} \quad M'Cl + e \rightarrow M' + Cl^-$$
$$\text{anode} \quad \tfrac{1}{2}M \rightarrow \tfrac{1}{2}M^{2+} + e$$
$$\text{overall} \quad \tfrac{1}{2}M + M'Cl \rightarrow \tfrac{1}{2}M^{2+} + M' + Cl^-$$

Characteristics of these batteries are given in Table 28.

Table 28
Some common sea water batteries

Battery	Operating temperature /°C	Voltage /V	Energy density /W h kg^{-1}	Energy density /W h dm^{-3}	Maximum power/kW
Mg|AgCl	−20 to +60	1–250	30–120	40–250	150
Zn|AgCl	−30 to +60	1–50	15–60	20–150	2.5
Mg|CuCl	−20 to +60	1–100	20–90	18–150	0.05

254 Secondary battery

Secondary battery. *See* **Electrochemical storage; Lead–acid battery.**

Secondary electrode. *See* **Three-electrode system.**

Sedimentation potential. *See* **Dorn effect.**

Siemens (S; units: $\Omega^{-1} = kg^{-1}\, m^{-2}\, s^3\, A^2$, $A\, V^{-1}$). The SI-derived unit of electrical conductance, that is reciprocal resistance. *See also* **Ohm**.

Sign convention. The IUPAC or Stockholm convention of e.m.f. values and electrode potentials in all forms of **reversible galvanic cell**, used throughout this book, can be summarized as follows.

1) The process of reduction involves a gain of electrons and oxidation a loss of electrons

$$\text{ox} + ne \underset{\text{oxidation}}{\overset{\text{reduction}}{\rightleftharpoons}} \text{red}$$

For example

$$Cu^{2+} + 2e \rightleftharpoons Cu$$
$$Fe^{3+} + e \rightleftharpoons Fe^{2+}$$

2) The sign of the **electrode potential**, on open circuit, is that which the electrode has with respect to the solution; that is, for the electrode $Cu^{2+} \mid Cu$ if there is a tendency for copper (II) ions to be discharged, then the electrode will become positively charged with respect to the solution, so the electrode is positive corresponding to the reaction

$$Cu^{2+} + 2e \rightarrow Cu$$

Thus a positive potential corresponds to a reduction process; electrode potentials on this convention are thus really reduction potentials.

3) The **hydrogen electrode**, with hydrogen gas at unit fugacity (1 atmos) in a solution of hydrogen ions of unit activity, is arbitrarily taken as the zero of electrode potentials at all temperatures, that is for the **half cell**

$$\begin{array}{c|c} H^+ & H_2(g), Pt \\ (a_{H^+} = 1) & (101\,325\ N\ m^{-2}) \end{array} \qquad E^{\ominus}_{H^+,H2} = 0$$

4) The half cells representing the reduction electrode potentials on open circuit written

$$\left.\begin{array}{l} Zn^{2+} \mid Zn \\ Cl^- \mid Cl_2, Pt \\ Cl^- \mid AgCl, Ag \\ Fe^{2+}, Fe^{3+} \mid Pt \end{array}\right\} \text{imply that the electrode processes are reduction} \left\{\begin{array}{ll} Zn^{2+} + 2e & \rightarrow Zn(s) \\ \tfrac{1}{2}Cl_2 + e & \rightarrow Cl^- \\ AgCl(s) + e & \rightarrow Ag(s) + Cl^- \\ Fe^{3+} + e & \rightarrow Fe^{2+} \end{array}\right.$$

Sign convention

and further that the potentials of these electrodes are the potential differences for the cells (in which the liquid junction potentials are ignored):

$$\left.\begin{array}{l} \text{Pt},H_2|H^+ \vdots Zn^{2+}|Zn \\ \text{Pt},H_2|H^+ \vdots Cl^-,Cl_2|\text{Pt} \\ \text{Pt},H_2|H^+ \vdots Cl^- |AgCl,Ag \\ \text{Pt},H_2|H^+ \vdots Fe^{2+},Fe^{3+}|\text{Pt} \end{array}\right\} \text{implying the reactions}$$

	$E^{\ominus}_{cell}$ /V
$\frac{1}{2}H_2 + \frac{1}{2}Zn^{2+} \rightarrow \frac{1}{2}Zn(s) + H^+$	-0.761
$\frac{1}{2}H_2 + \frac{1}{2}Cl_2 \rightarrow H^+ + Cl^-$	$+1.358$
$\frac{1}{2}H_2 + AgCl(s) \rightarrow Ag(s) + H^+ + Cl^-$	$+0.2225$
$\frac{1}{2}H_2 + Fe^{3+} \rightarrow Fe^{2+} + H^+$	$+0.783$

5) The sign of E_{cell} is the sign of the right-hand electrode (RHE): for example, for the cell

$$\begin{array}{c|c|c} \text{Pt},H_2(g) & H^+ & M^{z+} | M \\ (101\,325\ N\ m^{-2}) & (a_{H+} = 1) & \end{array}$$

$$E_{cell} = E_{RHE} - E_{LHE}$$
$$= E_{Mz+,M} - E_{H+,H2} = E_{Mz+,M} - 0$$

If E_{cell} is positive, then: (a) the process at the right-hand electrode is reduction; (b) the overall cell process is spontaneous from left to right, since $\Delta G = -nFE$, when $E > 0$, $\Delta G < 0$; (c) positive ions move from left to right through the cell, while negative ions move in the reverse direction, and electrons pass from left to right through the external circuit.

In the first of the above cells, comprising the standard hydrogen and zinc electrodes, the spontaneous cell reaction is, in fact, the reverse. So to conform with the experimental fact that the hydrogen electrode is more positive than the zinc, the cell should be written

$$\ominus\ Zn\ |\ Zn^{2+}\ \vdots\ H^+\ |\ H_2(g),Pt\ \oplus$$

for which $E^{\ominus}_{cell} = 0.761$ V and hence $E^{\ominus}_{Zn2+,Zn} = -0.761$ V for which the spontaneous cell reaction is

$$2H^+ + Zn(s) \rightarrow Zn^{2+} + H_2(g)$$

For the reaction

$$H_2(g) + AgI(s) \rightarrow Ag(s) + HI$$

ΔG changes sign when the molality of hydrogen iodide is about 0.1 so that the cell

$$\begin{array}{c|c} H_2(g),Pt\ |\ HI\ |\ AgI(s),Ag \\ (m) \end{array}$$

changes its polarity at about this concentration. The e.m.f. of the cell (as written) is positive for $m < 0.1$ and negative for $m > 0.1$ mol kg^{-1}.

6) The equation representing the electrode potential is

$$\Delta \phi = \Delta \phi^\ominus + \frac{RT}{nF} \ln \frac{a_{ox}}{a_{red}}$$

The equation for the e.m.f. of a chemical or **reversible galvanic cell** is

$$E_{cell} = \Delta \phi_{RHE} - \Delta \phi_{LHE} = \Delta \phi^\ominus_{RHE} - \Delta \phi^\ominus_{LHE} - \frac{RT}{nF} \ln \frac{\Pi a_{products}}{\Pi a_{reactants}}$$

$$= \frac{RT}{nF} \ln K_{therm} - \frac{RT}{nF} \ln \frac{\Pi a_{products}}{\Pi a_{reactants}}$$

7) When a **liquid junction potential** is included, E_j is added when E_{cell} is positive and subtracted when E_{cell} is negative.

Silver coulometer. *See* **Coulometer.**

Silver electrode. An electrode, consisting of metallic silver in contact with silver ions in solution and reversible to silver ions. The **electrode potential** is given by

$$E_{Ag+,Ag} = E^\ominus_{Ag+,Ag} + \frac{RT}{F} \ln a_{Ag+}$$

The standard electrode potential $E^\ominus_{Ag+,Ag} = 0.7991$ V. For accurate work, the silver electrode is coated with a fresh film of electrolytically deposited silver. See Ives and Janz (1961); James and Prichard (1977).

Silver–silver chloride electrode. A **reference electrode** for use in aqueous and non-aqueous solutions. It consists of a strip or disc of silver, on which is deposited a film of silver chloride. It behaves as a reversible chloride electrode with a potential given by

$$E_{AgCl,Ag,Cl-} = E^\ominus_{AgCl,Ag,Cl-} - \frac{RT}{F} \ln a_{Cl-}$$

The standard electrode potential is given by

$$E^\ominus_{AgCl,Ag,Cl-} = E^\ominus_{Ag+,Ag} + \frac{RT}{F} \ln K_{s,AgCl}$$

The electrode is prepared by anodizing a freshly plated silver electrode in hydrochloric acid until an even deposit of silver chloride is formed; this is unaffected by sunlight. The electrode potential is reproducible to ± 0.02 mV.

The standard electrode potential in aqueous solution at T K is given by

$$E^{\ominus}_{AgCl,Ag,Cl^-} = 0.22239 - 645.52 \times 10^{-6}(T - 298)$$
$$- 3.284 \times 10^{-6}(T - 298)^2 + 9.948 \times 10^{-9}(T - 298)^3$$

The standard electrode potential of silver–silver chloride electrodes in a range of solvents (*see* Appendix, Table 4) has been obtained from measurements of the e.m.f. of the cell

$$Pt,H_2(g) \mid HCl\text{-}S \mid AgCl,Ag$$

Traces of moisture in the solvent result in an increase in the e.m.f. of the cell; the wide variations reported in the literature are probably due to the use of non-anhydrous solvents.

The silver–silver chloride and the calomel electrodes cannot be used in liquid ammonia because the salts are either soluble or react with the solvent; the thallium amalgam–thallium chloride electrode has been used as a reference electrode in liquid ammonia.

Silver–silver bromide and silver–silver iodide electrodes prepared in a similar manner behave as reversible bromide and iodide electrodes, respectively. *See also* **Reversible galvanic cell.** See Ives and Janz (1961), chapter 4.

Silver–zinc battery. A lightweight battery having a silver oxide cathode and a zinc anode

cathode	$Ag_2O + H_2O + 2e \rightarrow 2Ag + 2OH^-$
anode	$Zn + 2OH^- \rightarrow Zn(OH)_2 + 2e$
overall	$Ag_2O + Zn + H_2O \rightarrow 2Ag + Zn(OH)_2$

A voltage of 1.8 V (which falls to 1.5 V) is achieved with a long discharge time and high capacity (70–120 W h kg^{-1}). Despite the initial cost, the silver–zinc battery is one of the best batteries currently available. The similar silver–cadmium cell gives 1.4–1.1 V at 48–75 W h kg^{-1}.

Sodium electrometallurgy. It is possible to discharge sodium electrolytically from an aqueous solution if a mercury cathode is used. On the basis of the standard electrode potentials ($E^{\ominus}_{Na,Na^+} = -2.71$ V), the decomposition of water to form hydrogen at the cathode would be a far easier process even when allowance is made for the high hydrogen **overpotential** (*see* **Hydrogen electrode reactions**) at mercury. However, the fact that sodium forms intermetallic compounds with mercury which are soluble in mercury and diffuse away so reduces the activity of sodium at the cathode surface, and its tendency to reionize, that the discharge of sodium becomes the preferred process. This method is not used commercially because of the high cost of extraction of sodium from its amalgam, and recourse is made to fused electrolytes. Sodium is manufactured by the electrolysis of the melt sodium chloride (42 wt percent), calcium chloride (58 wt percent) at 580°C in a cell having groups of concentric graphite anodes and steel cathodes, separated by metal diaphragms (a Down's cell). Sodium, being less dense than the melt, rises to the top of the cell where it is collected. Calcium is also formed by electrolysis but is precipitated by cooling the collected metal to 110°C when some 0.04 percent calcium remains.

Sodium–sulphur battery. A high-temperature (300–400°C) battery with a high capacity. Sodium reacts with sulphur to give sodium polysulphide

$$
\begin{array}{ll}
\text{cathode} & nS + 2e \rightarrow S_n^{2-} \\
\text{anode} & 2Na \rightarrow 2Na^+ + 2e \\
\hline
\text{overall} & 2Na + nS \rightarrow Na_2S_n
\end{array}
$$

The electrolyte is sodium β-alumina, which has a high electronic resistance but which allows the passage of sodium ions between the liquid electrodes (*see* **Solid electrolytes**). The cell voltage is 2.08 V and the theoretical energy density is 750 W h kg^{-1} (compare with **lead–acid battery**, which has a theoretical energy density of 170 W h kg^{-1}).

Solid electrolytes. Compounds that exhibit high conductivity ($\kappa \geq 10^{-1}\ \Omega^{-1}$ cm^{-1}) in the solid state with a low activation energy for conduction (around 0.1 eV) (i.e. superionic conductors). Examples of superionic conductors which have been studied are given in Table 29. A major use of these materials is as the electrolyte in all solid-state batteries. High **energy density** and high power density (*see* **Electrochemical storage**) coupled with good mechanical stability is possible, but in order to achieve this, high temperatures (250–500°C) are necessary.

There are three modes of conduction shown by solid electrolytes.

1) *Point defect conduction.* Defects may be produced thermally or by doping (compare this behaviour with electronic semiconductors). Silver chloride doped with cadmium creates silver (I) ion vacancies and leads to conductance by the migration of silver (I) ions.

2) *Conduction with 'internal melting'.* One ion remains fixed in the crystalline lattice (e.g., iodide ion in α-silver iodide) but the other ion is randomly distributed over a large number of interstitial positions. Such compounds are capable of good ionic conductivity, even at low temperatures. For example, RbAg$_4$I$_5$ has a conductivity of 0.25 Ω^{-1} cm^{-1} at room temperature.

3) *Ion-exchange conduction.* Sodium β-alumina (Na$_2$O . 11Al$_2$O$_3$) has a hexa-

Table 29
Properties of some common solid electrolytes

Solid electrolyte	Lattice type	Conductivity /Ω^{-1} cm^{-1}		Mobile ion
		25°C	500°C	
AgCl	Sodium chloride	$\leqslant 10^{-6}$	10^{-1}	Ag$^+$
β-AgI	Wurtzite	10^{-6}		Ag$^+$
LiI	Sodium chloride	$<10^{-6}$	$>10^{-2}$	Li$^+$
CaF$_2$	Fluorite		$>10^{-3}$	F$^-$
ZrO$_2$	Fluorite		10^{-2} (at 700°C)	O^{2-}
α-AgI	Anions in *bcc* lattice, cations in molten sublattice		10^0	Ag$^+$
RbAg$_4$I$_5$	Anions in *bcc* lattice, cations in molten sublattice	$>10^{-1}$		Ag$^+$
β-alumina Na$_2$O.11Al$_2$O$_3$	Spinel	10^{-1}	$>10^{-1}$	Na$^+$

gonal structure of the spinel block-type with all the sodium ions in a plane perpendicular to the *c*-axis of the spinel blocks. The sodium ion is highly mobile and may be exchanged for other alkali metal ions. This is in contrast to the behaviour of conductors of types (1) and (2). Batteries which employ solid electrolytes include lithium/lithium iodide/silver iodide (or copper (I) iodide) and the **sodium–sulphur battery** with sodium β-alumina as electrolyte. Other uses, or potential uses, of superionic conductors are in electrochromic devices, gas sensors, thermometers and thermoelectric generators.

Further reading
S. Chandra, *Superionic Solids*, North Holland Publishing, Amsterdam (1981)
P. Hagenmuller and W. Van Gool, *Solid Electrolytes*, Academic Press, New York (1978)

Solubility product (K_s; units: mol$^\nu$ kg$^{-\nu}$). For a sparingly soluble salt which dissociates into $\nu+$ positive ions and $\nu-$ negative ions, the solubility product, or activity solubility product, is defined by

$$K_s = a_{z+}^{\nu_+} a_{z-}^{\nu_-} \tag{113}$$

Eq. (113) applies to any saturated solution, but is only of any practical interest

260 Solubility product

for sparingly soluble salts. For very dilute solutions, the **activity coefficient** can be taken as unity and hence Eq. (113) can be written as

$$K_s = m_+^{\nu_+} m_-^{\nu_-} \tag{114}$$

For a 1:1 sparingly soluble salt, $m_+ = m_- = K_s^{1/2}$. So long as pure solid is in equilibrium with the dissolved solute, any modification which causes a change in the solubility ($m_\pm$), such as the addition of an electrolyte must cause a change in the activity coefficient since at constant temperature and pressure the solubility product is constant

$$K_s = a_+ a_- = a_\pm^2 = m_\pm^2 \gamma_\pm^2 \tag{115}$$

If the added electrolyte contains an ion in common with the sparingly soluble salt, there is a decrease in the solubility of the sparingly soluble salt, whereas if the added electrolyte contains no common ion, there is an increase in the solubility of the sparingly soluble salt.

Determination of the solubility product.
1) From solubility measurements in the presence of added electrolytes. Eq. (115) can be written in the form

$$\log \gamma_\pm = \log K_s^{1/2} - \log m_\pm$$

If $m_\pm$ is measured in the presence of known amounts of added electrolyte, the graph of $\log m_\pm$ against $I^{1/2}$ is of intercept $\log K_s^{1/2}$, whence the solubility product and hence the activity coefficient at any concentration can be obtained.

2) From e.m.f. measurements using a cell without a liquid junction. For the determination of the solubility product of silver chloride a suitable cell would be

$$\ominus \quad Ag, AgCl(s) \mid HCl(aq) \mid Cl_2(g), Pt \quad \oplus$$

for which

$$\ln K_s = (E^\ominus_{Cl_2, Cl^-} - E^\ominus_{Ag^+, Ag} - E_{cell}) \frac{F}{RT}$$

3) From e.m.f. measurements using a **concentration cell** without a liquid junction. For the determination of the solubility product of silver chloride a suitable cell would be

| Ag | AgCl(s) KCl (0.01 mol dm^{-3}) ($a_{Ag^+, 1}$) | KNO$_3$ saturated solution | AgNO$_3$ (0.01 mol dm^{-3}) ($a_{Ag^+, 2}$) | Ag |

in which the solution in the left-hand compartment is obtained by adding one drop of a silver nitrate solution to the potassium chloride solution. The e.m.f. is given by

$$E = \frac{RT}{F} \ln \frac{a_{Ag+,2}}{a_{Ag+,1}}$$

From a knowledge of the activity coefficient of silver ions in 0.01 mol dm^{-3} silver nitrate solution, the activity of silver ions in the potassium chloride can be determined. From this and the activity coefficient of chloride ions in the potassium chloride solution the solubility product can be calculated.

4) From measurements of the **conductivity** of a saturated solution of the sparingly soluble salt and the known individual **molar ionic conductivity** values at infinite dilution, the concentration of the salt can be determined thus

$$c = \frac{\kappa}{\Lambda^o_+ + \Lambda^o_-}$$

and hence the solubility product. See James and Prichard (1977).

Solvated electron. A chemical entity which plays a fundamental role in a wide range of electron transfer reactions, and which can provide a sensitive probe into condensed states of matter.

Solvated electrons are created by irradiation by X- or γ-rays, electrochemical injection, or, in the case of ammonia, the dissolution of an alkali metal.

Inert solvents such as liquid krypton, argon, methane or tetramethylsulphoxide lead to the thermalization of the electron in a quasi-free state. In water, ammonia or certain glasses, the electron may be trapped or solvated. A trapped electron exists in a potential well in a solvent that does not change its structure to any great extent. A trapped electron may give way to a solvated electron if the solvent reorganizes around the electron site. Figure 111 shows the structure of a solvated electron in 10 mol dm^{-3} sodium hydroxide glass determined by spin echo electron spin resonance spectroscopy.

Further reading
B. C. Webster, Electron solvation phenomena, *Annual Reports C*, Royal Society of Chemistry, London (1979)

Solvation of ions. The changes accompanying the introduction of an ion into a solvent. If the solvent is water then the process is known as hydration.

Consideration of the structural modification imposed on the solvent by the ion leads to two approaches. On the one hand, one can regard the solvent as being a homogeneous continuum where any change can be considered to be smoothly and continuously dissipated from the site of solvation. On the other hand, on a more structural basis, a microscopic viewpoint may be adopted

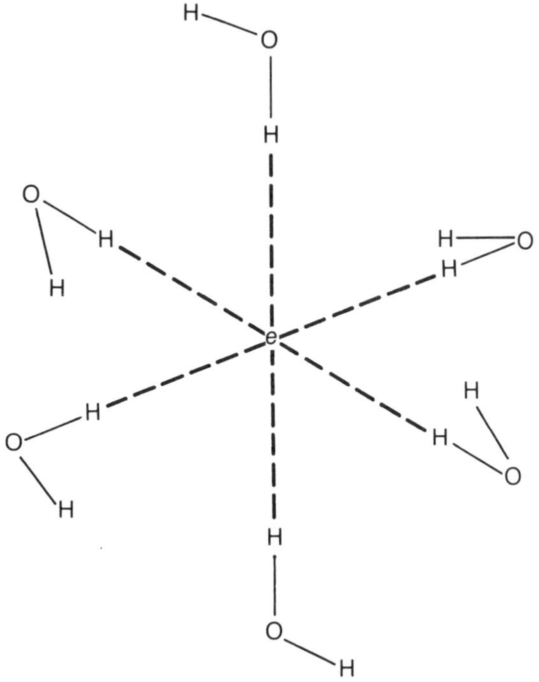

Figure 111. Structure of a solvated electron in a sodium hydroxide aqueous glass showing the octahedral coordination of the electron. The cavity formed by the water molecules is 0.42 nm across

which considers the reorientation of the solvent molecules around an ion. The first of the continuous models of ionic solvation is due to Born who considered the thermodynamics of introducing a neutral species from infinity into the solution, charging up the ion to its correct valency followed by the relaxation of the surrounding solvent.

The structure of the solvent in the immediate vicinity of the ion is discussed in terms of solvent shells. Three regions are recognized around a solute ion. Immediately next to the ion, solvent molecules of the primary solvation shell are considered to have a greater energy of attraction to the ion than to each other in the bulk solution. A secondary region may occur between the strongly bound solvent molecules of the primary shell and the bulk solvent. This region may be one of order, in which the attraction of the central ion is sufficient to orientate outer solvent layers, or a disordered region (particularly around larger ions) in which the ion is less efficient at reorientating the free solvent molecules. Solutes are thus classed as structure makers or structure breakers in terms of their effect on the secondary solvation shell.

The number of solvent molecules in each region is known as the primary solvation number and the secondary solvation number. The values of these quantities depends on the method of determination, but it is usually found that cations are more solvated than anions, and that smaller, highly charged ions are more solvated than larger ions of lower charge. *See also* **Ion pair; Molar ionic conductivity**.

Solvent correction. *See* **Conductivity**.

Spectroelectrochemistry. The use of spectroscopic methods to study electrochemical reactions.

Optical spectroelectrochemistry.
1) Transmission methods (*see* Figure 112(a)). A beam of light passed through an **optically transparent electrode** (OTE) is used to monitor the concentration of reactants or products as an electrochemical reaction proceeds: for example, in a reaction yielding a product which is allowed to diffuse away from the electrode, the absorbance (A) is related to the diffusion coefficient and time by

$$A = \text{const } D^{1/2} t^{1/2}$$

The diffusion coefficient is obtained from a plot of the absorbance against the square root of the time. Transmission spectroelectrochemistry, while having the advantages of non-interacting, *in situ* methods, is limited in wavelength to the window afforded by the solvent and the OTE.

2) Reflectance methods (*see* Figure 112(b)). Of particular use in observing changes during film growth on bright metal electrodes, specular reflectance and **ellipsometry** have provided important data on the mechanisms of **corrosion**. Reflectance experiments commonly measure the change of intensity and, in ellipsometry, the phase of polarized light, as the potential at an electrode is varied.

3) Internal reflection. By careful manipulation of the path of a light beam through an OTE, a series of total internal reflections may be obtained. The passage of light many times through the surface layers augments the final signal that is detected and allows a greater sensitivity in monitoring the electrode surface.

4) Photothermal methods. An intense beam of light falling on an electrode produces temperature changes which affect the electrochemical reaction. The beam is chopped and the resulting modulation in the electrochemical parameter measured.

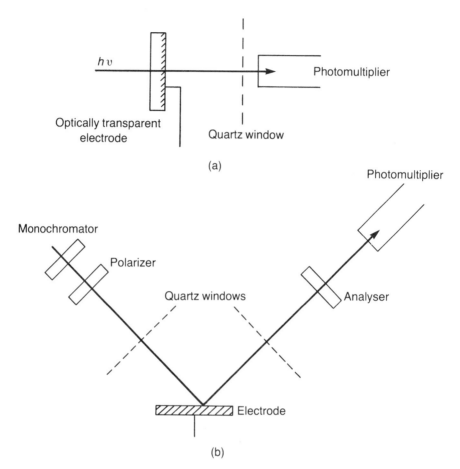

Figure 112. Optical arrangements of optical spectroelectrochemical experiments: (a) transmission; (b) reflectance

5) Raman spectroscopy. The measurement of inelastically scattered radiation from an electrode surface. The modulation of a beam of monochromatic radiation (ultraviolet or visible from a laser) gives vibrational information which, because of the absorption of the solvent, could not be obtained by infrared spectroscopy. The signals from certain surfaces are increased by factors of 10^4–10^6 in resonance Raman spectroscopy. The existence of levels in the material which coincide with the energy of the excited states produced by the beam allows a more efficient transfer of energy to the excited states and thus an enhanced signal.

Photoelectron spectroscopy (PES). The measurement of the energy of electrons ejected from a surface on bombardment with X-rays (X-ray photoelectron spectroscopy (XPS) or electron spectroscopy for chemical analysis (ESCA)), ultraviolet (ultraviolet photoelectron spectroscopy (UPS)) or electrons (Auger electron spectroscopy (AES)) is an extremely powerful analytical tool in the study of surfaces. The penetration of the radiation is limited to a few atomic layers and so information is obtained only from the surface itself, a fact which recommends these methods to electrochemists. In order to detect and measure the emitted electrons, all PES methods require high vacua. In the study of electrochemistry, therefore, the electrode must be transferred from the electrochemical cell to the spectrometer as quickly as possible with the least chance of contamination. This takes of the order of minutes and thus short-lived species cannot be studied by these methods.

Electron spin resonance spectroscopy (esr). A technique of magnetic resonance which observes unpaired electron spin in radicals and atoms. The radical anion $PhNO_2^{-\cdot}$ generated by the electrochemical reduction of nitrobenzene (*see* **Organic electrochemistry**)

$$PhNO_2 + e \rightarrow PhNO_2^{-\cdot}$$

was an early example of the detection of an intermediate radical by esr spectroscopy. Because of the short-lived nature of radicals, the cell may be constructed in the cavity of the spectrometer. Even so it is not possible to detect very reactive radicals. The technique of 'spin trapping' in which a molecule is introduced which can scavenge a radical to give a longer-lived second radical has been tried as one solution to the problem of detecting reactive intermediates. Nitrones or nitroso compounds have been used as spin traps

$$\begin{array}{c} R' \\ \diagdown \\ R'' \end{array}\!\!C\!=\!\overset{+}{N}\!\!\begin{array}{c} \diagup O^- \\ \\ \diagdown R'' \end{array} + R^{\cdot} \rightarrow \begin{array}{c} R' \\ \diagdown \\ R'' \end{array}\!\!C\!-\!N\!\!\begin{array}{c} \diagup O^{\cdot} \\ \\ \diagdown R'' \end{array}$$

$$R'\!-\!N\!=\!O + R^{\cdot} \rightarrow \begin{array}{c} R' \\ \diagdown \\ R'' \end{array}\!\!N\!-\!O^{\cdot}$$

The interpretation of the esr spectrum produced, particularly with nitrones which trap the radical two atoms away from the unpaired electron, may be difficult. Also the potentials in the system must be within the limits of stability of the spin trap itself. However, these methods have shown promise in the determination of mechanisms of organic electrochemical reactions. Work with semiconducting electrodes has shown that light may interact with the system producing photoelectrochemical reactions (*see* **Photoelectrochemistry**). See Bard and Faulkner (1980).

Stability constant. The reciprocal of the **formation constant**.

Standard cell. A cell used for purposes of comparison and in the calibration of measuring circuits. A standard cell should provide a constant and accurately reproducible e.m.f. and should not deteriorate on standing. The Weston cell

$$\ominus \; Cd(Hg) \;|\; CdSO_4 \cdot \tfrac{2}{3}H_2O(s) \;|\; \begin{array}{c}\text{saturated solution}\\ CdSO_4\end{array} \;|\; \begin{array}{c}Hg_2SO_4\\ \text{paste}\end{array} \;|\; Hg \; \oplus$$

has a low temperature coefficient; at T K the e.m.f. is given by

$$E_T = 1.018\,30 - 4.06 \times 10^{-5}(T-293) - 9.5 \times 10^{-7}(T-293)^2$$

The Clark cell

$$\ominus \; Zn(Hg) \;|\; ZnSO_4 \cdot 7H_2O(s) \;|\; \begin{array}{c}\text{saturated solution}\\ ZnSO_4\end{array} \;|\; \begin{array}{c}Hg_2SO_4\\ \text{paste}\end{array} \;|\; Hg \; \oplus$$

has a higher temperature coefficient; at T K the e.m.f. is given by

$$E_T = 1.433\,0[1 - 8.4 \times 10^{-3}(T-288)]$$

Standard electrode potential. *See* **Electrode potential.**

Streaming potential. One of the electrokinetic phenomena (*see* **Electrokinesis**). When a liquid is forced through a capillary tube or through a plug of finely divided material, a potential difference arises between the two ends of the material (*see* Figure 113). The hydrostatic pressure is measured by the difference in level of the liquid in the two reservoirs. The potential difference is measured between the two platinum electrodes which enclose the material under investigation.

The effect is a consequence of the **electrical double layer** which exists at an interface. The ζ-**potential** at the surface may be calculated from the equation

$$E = \frac{\zeta \varepsilon P}{4\pi \eta \kappa}$$

where E is the measured potential difference and P is the hydrostatic pressure.

Stress corrosion cracking. Increased corrosion activity that occurs at a microcrack in a material under stress, which results in a rapid propagation of the crack.

At the apex of a crack in a metal corrosion will occur (*see* Figure 114), the supporting reaction taking place at the exposed surface; for a detailed discussion of why this should be, *see* **Corrosion**. A high current density is found for the metal dissolution in the small surface area of the crack, which has a tendency to gouge out and widen the crack, and thus normalize the current density. If, however, the metal is under stress, even if this is not sufficient to distort the bulk metal, plastic deformation along the crack occurs.

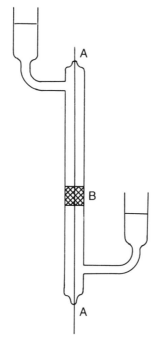

Figure 113. Streaming potential cell: A, platinum wire leads to platinum gauze electrodes; B, plug of material under study

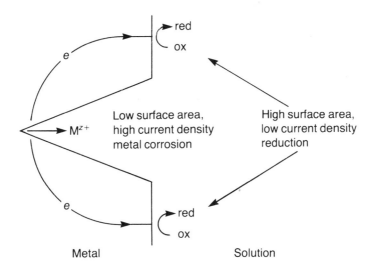

Figure 114. Representation of the corrosion of a metal in a crack at its surface

Stripping voltammetry

The crack is propagated, and the high current density corrosion continues as fresh metal surfaces are exposed. A piece of metal under stress may therefore appear to be corroding at a low, normal rate until the microcracks become numerous and join up, when a sudden failure of the structure occurs. The small surface cracks may be often initiated by **hydrogen embrittlement**.

Stripping voltammetry. A range of analytical techniques in which the substance to be analysed is first concentrated electrochemically at the surface of an electrode (or in a mercury amalgam) and then is stripped off, again electrochemically. The stripping process is monitored by following the current passed. Since the concentration of the species is much greater at the electrode than in the bulk of the solution, an increase in sensitivity is gained over methods which attempt to sense the species directly in solution. Voltammetric stripping methods are classed as anodic or cathodic, depending on the polarity of the electrode during stripping.

In stripping voltammetry, complete electrolysis of the substance from the solution may be performed, but more usually pre-electrolysis is carried out for a fixed time which concentrates a certain fraction (2–3 percent) of the substance.

Preconcentration.
1) Metals that form a sufficiently concentrated amalgam with mercury (bismuth, copper, thallium, lead, cadmium, antimony, indium, zinc) can be preconcentrated at a mercury electrode. Anodic dissolution of the metal from the amalgam yields the stripping current.
2) Metallic ions (silver, copper, mercury) may be reduced to the metal on an inert (platinum, gold, carbon) electrode. The metals are then stripped anodically.
3) Concentration may involve the formation of a sparingly soluble compound. The electrode may react, as in the case of the determination of chloride ions, on silver or mercury, or an added component of the electrolyte may yield the compound.
4) Surface complexes may be produced with a suitable agent adsorbed on the electrode surface or present in the solution.

Stripping. The stripping process may be performed potentiostatically or galvanostatically by a steady-state method or by variation of the controlled parameter (i.e. potential or current). *See also* **Linear sweep voltammetry; Polarography; Transient methods**.

Further reading
F. Vydra, K. Stalik and E. Julakova, *Electrochemical Stripping Analysis*, Ellis Horwood, Chichester (1976)

Sulphur dioxide probe. A **gas-sensing membrane probe** which measures the partial pressure of sulphur dioxide in solution; sulphite, bisulphite and metabisulphite concentrations can be measured after the solution has been acidified.

The probe is virtually insensitive to gradual temperature variations in the range 0–40°C. Calibration with solutions containing known concentrations of sulphur dioxide is necessary; the response is linear from 3–3000 mg dm^{-3} sulphur dioxide. Acidic species (in particular acetic acid) interfere with this probe.

Superionic conductors. *See* **Solid electrolytes.**

Surface area determination. Electrochemical methods for the determination of surface area rely on measuring a property of the electrode which depends on the amount of electrochemically active surface.

The capacity of the double layer (*see* **Electrical double layer**) in the non-faradaic region (i.e. in the absence of an electrochemical reaction and, hence, **pseudocapacitance**) may be compared with that at a smooth mercury electrode. A value of about 17 μF cm^{-2} is found for mercury in many aqueous electrolytes and thus greater values for rough metal electrodes are interpreted in terms of increased surface area.

The electro-oxidation or reduction of adsorbed species can give an accurate determination of surface area, if the area taken up by a single adsorbed molecule is known. The adsorption of hydrogen as atoms on metals on which a monolayer is formed (e.g., platinum) is an example of this technique. Hydrogen is bubbled over the electrode surface, which is held at near 0 V versus standard hydrogen electrode. The surface becomes covered with hydrogen atoms which are stable at this potential. Molecular hydrogen is then purged away with nitrogen, and the adsorbed atoms are oxidized, the charge passed being a measure of the number of atoms. Oxidation may be accomplished galvanostatically (constant current), when the **transition time** is measured, or potentiostatically, when the potential is pulsed anodically and the burst of current measured with time (*see* **Transient methods**). The measurement of oxide formation or reduction may also give values of the surface area.

Surface charge. The charge at the surface of a material (solid, colloidal particle, biological cell) in contact with a polar medium may be due to adsorption of ions at the surface—a non-ionogenic surface (e.g., chloride ions on a silver chloride sol, on hydrocarbon oil droplets or on air bubbles)—or to the ionization of groups at the surface—an ionogenic surface (e.g., sulphonate groups on a sulphonated polystyrene latex particle, amino and carboxyl groups from proteins on biological cells).

270 Surface charge

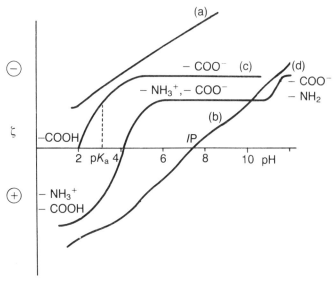

Figure 115. Typical curves for variation of ζ-potential with pH: (a) non-ionogenic surface; (b) protein; (c) carboxyl colloid; (d) biological cell with surface amino acid carboxyl groups

The charge on a non-ionogenic surface is usually negative because anions are smaller, less hydrated and more polarizing than cations. The ζ-**potential** becomes more negative with increasing pH due to the adsorption of hydroxyl ions and the desorption of protons (*see* Figure 115, curve a).

The charge carried by proteins, due to the ionization of carboxyl and amino groups, depends on the pH of the medium (*see* Figure 115, curve b); at pH values below (above) the **isoelectric point**, the protein is positively (negatively) charged. Carboxyl colloids (e.g., mannuronic acid-coated oil droplets) and bacteria with a carboxyl-type capsule have zero charge in suspension at low pH values (*see* Figure 115, curve c). This becomes more negative with increase in the pH of the suspension medium and attains a constant value due to the dissociation of the carboxyl groups. The pH at which the ζ-potential is half the maximum constant value gives a value for the pK_a of the surface groups; this can be of importance in identifying the surface ionogenic groups. The pH–ζ-potential curve for biological cells with both surface carboxyl and amino groups shows a series of steps due to the ionization of different surface species at different pH values (*see* Figure 115, curve d). Chemical and enzymic treatments of such surfaces brings about marked changes in the position and the shape of the pH–ζ-potential curves; such changes have revealed the nature and quantity of surface-charged groups.

Further reading
A. M. James, Electrophoresis of particles in solution, *Surface Colloid Sci.*, **11**, 121 (1979)

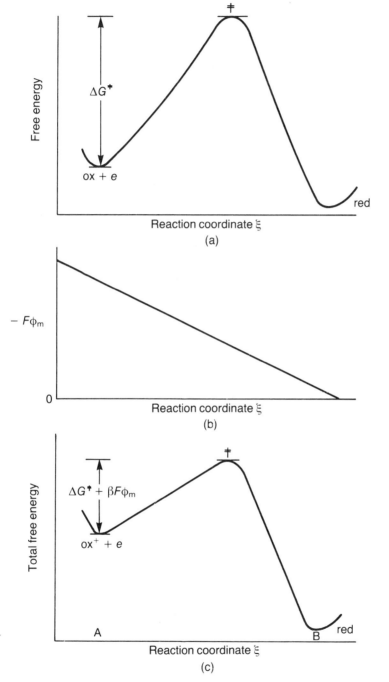

Figure 116. The effect of an applied electric field on the free energy profile of a reaction ox$^+$ + e → red. β = A‡/AB. (c) = (a) + (b)

272 Surface potential

Surface potential (dipole or chi, χ, potential). A potential associated with the work done on crossing a surface dipole layer. An adsorbed layer on an electrode creates a solution surface potential (χ_S). The surface potential of a metal (χ_M) arises from the finite chance of finding an electron outside the metal surface. Thus across an electrode/electrolyte interface, a surface potential difference is defined as

$$\Delta\chi = \chi_M - \chi_S$$

Symmetry coefficient (β). The concept of the symmetry coefficient was introduced into the derivation of the **Butler–Volmer equation** to allow for the fact that only a fraction of the potential drop between the pre-electrode state and the electrode surface contributes to the activation energy. Figure 116 shows the free energy profile in the absence of potential, and with $F\phi_m$ at the electrode. It may be seen that in passing along the reaction coordinate the field lowers the height of the transition state by $\beta F\phi_m$, where

$$\beta = \frac{\text{distance along the reaction coordinate to } \ddagger}{\text{total distance along the reaction coordinate}}$$

This ratio may be expressed in terms of the shapes of the Morse curves which give rise to the form of the free energy profile in Figure 116 (*see* p. 271).

For simplicity consider a one-electron transfer resulting in the formation of a neutral molecule (e.g., $Ag^+ + e \rightarrow Ag$). Figure 117 shows the effect of a

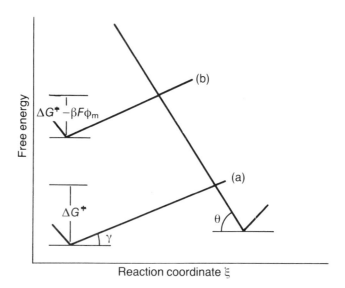

Figure 117. Linearized Morse curves of reactants and products of an electrochemical reaction: (a) in absence of applied potential; (b) ϕ_m at electrode

potential ϕ_m on the energy of the system which is drawn as having linear free energy profiles. The point of intersection of the free energy surfaces for oxidation and reduction moves as the potential is applied. Another way of looking at the symmetry factor is

$$\beta = \frac{\text{change of activation energy}}{\text{change of electrical energy}}$$

$$\beta = \frac{\tan \gamma}{\tan \gamma + \tan \theta}$$

The importance of this latter view is that β is now related to the shapes of potential energy curves and hence to molecular properties. See Bockris and Reddy (1973).

T

Tafel equation. The high-field-limiting case of the **Butler–Volmer equation**.

The current density–overpotential relation of an activation-controlled one-electron electrodic reduction ox + e → red is

$$i = i_o \left\{ \exp\left(-\frac{\beta F}{RT}\eta\right) - \exp\left[\frac{(1-\beta)F}{RT}\eta\right] \right\} \quad (13)$$

If the **overpotential** η for this reduction is high (and negative), that is

$$\exp\left(-\frac{\beta F}{RT}\eta\right) \gg \exp\left[\frac{(1-\beta)F}{RT}\eta\right]$$

Eq. (13) reduces to

$$i \approx i_o \exp\left(-\frac{\beta F}{RT}\eta\right) \quad (116)$$

Eq. (116) is the Tafel equation and is usually valid for $\eta > 0.1$ V (*see* Figure 118). The Tafel region extends up to the point where diffusion is significant, and the current becomes limited by the rate of supply of reactants to the electrode surface (*see* **Diffusion-limited current**).

Taking logarithms of Eq. (116)

$$\ln i = \ln i_o - \frac{\beta F}{RT}\eta$$

or, for a multistep reaction

$$\ln i = \ln i_o - \frac{\alpha F}{RT}\eta$$

(*see* **Transfer coefficient**). A plot of $\ln i$ against η will give β (or α) from the slope and $\ln i_o$ as the intercept (*see* **Exchange current density**). As current or potential may be the dependent variable in an electrochemical experiment, η may be plotted against $\ln i$, or $\log i$; indeed the 'Tafel slope' is often quoted in mV per decade. At room temperature, $2.303\, RT/F$ has the value 60 mV and so the Tafel slope for a reaction having $\alpha = \frac{1}{2}$ would be 120 mV decade^{-1}. The Tafel slope may not be given by Eq. (116) when diffusion effects are important. This may be the case for a **porous electrode**; steady-state data may not be reliable and recourse must be had to **transient methods**. A diminution

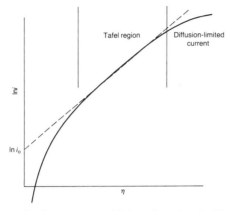

Figure 118. Current density–overpotential data plotted as the Tafel equation

of the Tafel slope may be due to double layer effects. This is only important in less concentrated solutions at potentials near the **potential of zero charge**.

Tafel reaction. *See* **Hydrogen electrode reactions.**

Teflon-bonded electrode. A **porous electrode** in which particles of Teflon (polytetrafluoroethylene, PTFE) provide dry channels for gases to diffuse through. Teflon also acts as an inert, mechanically strong support for small catalyst particles. A typical catalyst to Teflon ratio is 10:3 by weight. The catalyst powder is ultrasonically mixed with a Teflon dispersion in water containing 50 percent Teflon and a small amount of liquid (water or methanol). The resulting slurry is painted on to a conducting screen and, after drying in air, the electrode is cured at 300°C. The Teflon fuses to create a matrix which holds the catalyst particles in contact with the mesh which carries the current.

Thermodynamics of cells. If a **reversible galvanic cell** of e.m.f. E drives a perfectly reversible motor which delivers all the electrical energy it receives as work, then for n faradays of electricity passing through the cell and motor the work done is nFE joules. This is the maximum net work (excluding work due to change in volume) obtainable from the reaction which takes place in the cell; hence

$$w' = -\Delta G = nEF \qquad (117)$$

From a knowledge of the temperature coefficient $(\partial E/\partial T)_P$, ΔH can be calculated from the Gibbs–Helmholtz equation

$$\Delta H = \Delta G - T\left(\frac{\partial(\Delta G)}{\partial T}\right)_P = -nF\left[E - T\left(\frac{\partial E}{\partial T}\right)_P\right] \qquad (118)$$

Thermoneutral potential

The enthalpy change of a reaction, ΔH, can usually be determined more accurately from e.m.f. data than by direct calorimetry; the method is, however, limited to reactions which take place in a reversible cell. ΔS can also be obtained from the same data; the quantity

$$nFT\left(\frac{\partial E}{\partial T}\right)_P = -T\left(\frac{\partial(\Delta G)}{\partial T}\right)_P = T\Delta S \tag{119}$$

is often referred to as the reversible heat absorbed during the working of the cell. If $\partial E/\partial T$ is positive (negative), then heat is absorbed (evolved) in the working of the cell, and the electrical energy is greater (less) than the decrease in the enthalpy; when $\partial E/\partial T = 0$ then $\Delta H = \Delta G$. The thermodynamic **equilibrium constant** for a cell reaction can be determined from a knowledge of the standard e.m.f. of the cell

$$\ln K_{\text{therm}} = -\frac{\Delta G^\ominus}{RT} = \frac{nFE^\ominus}{RT} \tag{120}$$

From a knowledge of the temperature coefficient of e.m.f. of a galvanic cell the entropy of individual ions can be calculated. Thus the value $S^\ominus_{\text{Cl}^-}$ can be calculated from e.m.f. data at different temperatures for the cell

$$\ominus \quad \text{Pt,H}_2(\text{g}) \quad | \quad \text{H}^+\text{Cl}^- \quad | \quad \text{AgCl,Ag} \quad \oplus$$
$$(101\,325 \text{ N m}^{-2})$$

for which the cell reaction is

$$\tfrac{1}{2}\text{H}_2(\text{g}) + \text{AgCl}(\text{s}) \rightarrow \text{H}^+ + \text{Cl}^- + \text{Ag}(\text{s})$$

The value of $(\partial E/\partial T)_P = -6.4552 \times 10^{-4}$ V K^{-1}, thus from Eq. (119)

$$\Delta S^\ominus = -1 \times 96\,487 \times 6.4552 \times 10^{-4} = -62.28 \text{ J K}^{-1} \text{ mol}^{-1}$$

For the reaction, where $S^\ominus_{\text{H}^+} = 0$ by convention,

$$\Delta S^\ominus = S^\ominus_{\text{Ag}} + S^\ominus_{\text{H}^+} + S^\ominus_{\text{Cl}^-} - \tfrac{1}{2}S^\ominus_{\text{H}_2} - S^\ominus_{\text{AgCl}}$$

$$= 42.70 + 0 + S^\ominus_{\text{Cl}^-} - \frac{130.59}{2} - 96.10$$

whence $S^\ominus_{\text{Cl}^-} = 56.41$ J K^{-1} mol^{-1}. See James (1976).

Thermoneutral potential. The reversible thermodynamic potential of an electrochemical reaction is given by

$$E^\ominus = -\frac{\Delta G^\ominus}{nF} = -\frac{\Delta H^\ominus}{nF} + \frac{T\Delta S^\ominus}{nF}$$

$-\Delta H^\ominus/nF$ is the thermoneutral potential. The significance of $-\Delta H^\ominus/nF$ may be seen with reference to the **electrolysis of water**. For the reaction

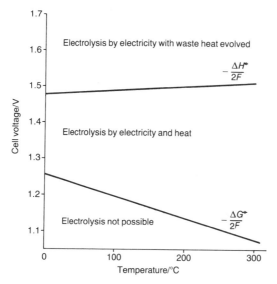

Figure 119. Temperature variation of the standard potential $-\Delta G^{\ominus}/2F$ and the thermoneutral potential $-\Delta H^{\ominus}/2F$ with temperature for the electrolysis of water

$H_2O \rightarrow H_2 + \tfrac{1}{2}O_2$ at 25°C, $\Delta G^{\ominus} = -246$ kJ mol^{-1}, $\Delta H^{\ominus} = -286$ kJ mol^{-1} and $\Delta S^{\ominus} = 164$ J K^{-1} mol^{-1}. The thermoneutral potential of this reaction is 1.47 V. The value of $T\Delta S^{\ominus}$ is the heat which must be supplied to the system from the surroundings during electrolysis. If this heat is available, electrolysis will occur at potentials greater than $E^{\ominus}$. The potential needs to be raised to the thermoneutral potential, however, if the reaction is to proceed isothermally. Figure 119 illustrates the variation in the standard and thermoneutral potential with temperature. If hydrogen is produced by electrolysis at potentials between the standard and thermoneutral potentials, and is then burnt to give heat, the overall efficiency of the cycle may reach 120 percent.

Three-electrode system. In any electrochemical cell, two electrodes at which the oxidation and reduction reactions take place are essential. The electrodes may be joined through a voltmeter for equilibrium measurements. Alternatively, the electrodes may be connected to a **potentiostat** or **galvanostat**, or be different sites on the same piece of material, as in **corrosion**. For measurements made of a dynamic system, that is one in which current flows, a third (reference) electrode is usually present in the cell. The need to be able to measure the potential of a given electrode against a reference may be appreciated by considering the voltage of a simple two-electrode cell. The electrode under investigation is known as the 'working' or 'test' electrode. The electrode which completes the circuit is the 'secondary' or 'auxiliary' electrode. The cell voltage, the potential between the working and secondary electrodes, is

$$E_{cell} = \phi_{WE,e} + \eta_{WE} - \phi_{SE,e} - \eta_{SE} + IR$$

(*see* **Inner potential; Overpotential**). The potential drop through the solution caused by the resistance of the electrolyte is given by IR, where I is the current flowing and R is the resistance. By a measurement of E_{cell} alone, it is impossible to know the overpotential of the working electrode independently of that of the secondary electrode. If a reference electrode is introduced into the cell, the potential between the working electrode and the reference electrode, measured with a high-impedance voltmeter which allows no current to flow between them, is

$$E_{WE-RE} = \phi_{WE,e} + \eta_{WE} - \phi_{RE,e} + IR_{WE-RE}$$

IR_{WE-RE} is the potential drop in the solution between working and reference electrodes caused by the flow of current in the cell through that resistance (R_{WE-RE}). The so-called IR drop is minimized by use of a **Luggin capillary** and may be measured by the **interruptor technique**. The difference between E_{WE-RE} measured when current flows and the value at equilibrium (measured or calculated) is

$$\begin{aligned}\Delta E &= \phi_{WE,e} + \eta_{WE} - \phi_{RE,e} + IR_{WE-RE} - (\phi_{WE,e} - \phi_{RE,e}) \\ &= \eta_{WE} + IR_{WE-RE}\end{aligned}$$

Hence the overpotential at the working electrode may be determined.

Experimentally a commonly used design of cell has three compartments for the working, secondary and reference electrodes (*see* Figure 120). If it is undesirable to allow mixing of reactants and products of one of the half-cell reactions, a glass frit may be incorporated between the secondary and working electrode compartments. The cell may also be provided with inlets and outlets for gases: for example, a reactant gas in a fuel cell or nitrogen to deoxygenate a solution. A high degree of purity of electrolyte and electrodes is required for careful work. Cells may therefore be designed to allow *in situ* distillation of solutions, the encapsulation of electrodes which are broken only when they are to be used and a completely inert atmosphere.

Throwing power. *See* **Electroplating.**

Tin plating. A method of protecting easily corroded metals by the deposition of a thin layer (about 3 μm) of tin. Hot plating in molten tin is widely used, but this method is wasteful of tin. Electroplating from alkaline solutions of tin (IV) or acidic solutions of tin (II) gives thinner coats which use the tin more efficiently. A maximum of 90 percent current efficiency is obtained from aqueous sulphate or halide solutions, but 100 percent efficiency and superior

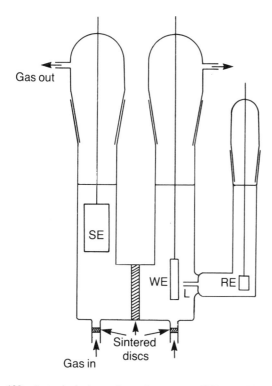

Figure 120. A typical three-electrode system. WE, working electrode; SE, secondary electrode; RE, reference electrode; L, Luggin capillary

plate quality is found for plating from the melt of 80 percent tin (II) chloride/20 percent potassium chloride. When tin anodes are used very little overpotential is required and the process occurs practically at the equilibrium potential. Tin may also be deposited as an alloy with other metals: for example, copper, lead, nickel and zinc (*see* **Alloy electrodeposition**).

Titration. *See* **Electrometric titrations.**

Transfer coefficient. Multistep electrochemical reactions, when controlled by the rate of one step, follow a current–overpotential equation which is similar to the **Butler–Volmer equation** (Eq. (13))

$$i = i_o \left[\exp\left(-\frac{\alpha_c F}{RT} \eta\right) - \exp\left(\frac{\alpha_a F}{RT} \eta\right) \right]$$

α_c and α_a are the cathodic and anodic transfer coefficients, respectively. Unlike the **symmetry coefficient** (β) for a single electron transfer, the transfer coefficient may take any positive value and is not confined between zero and one. Transfer coefficients are determined experimentally from Tafel plots (*see* **Tafel equation**), or the variation of i_o with concentration of the oxidized or reduced species.

The relationship of α_a and α_c to the mechanism of a reaction may be seen by considering the plating and dissolution of copper.

$$Cu^{2+} + e \rightleftharpoons Cu^+ \qquad (a)$$
$$Cu^+ + e \rightleftharpoons Cu \qquad (b)$$

There are two possible mechanisms, with either the first or second step being rate-limiting. If the reduction of copper (II) to copper (I) ions is rate-limiting, with copper (I) ions to copper as a fast step, the overall reaction will be given by the rate of the single electron transfer of reaction (a), and therefore the transfer coefficient will be equal to the symmetry coefficient for that reaction. Thus $\alpha_c = \beta \approx 0.5$. If reaction (b) is rate-limiting, the cathodic current density is

$$i = A a_{Cu+} \exp\left(-\frac{\beta F}{RT}\eta\right) \qquad (121)$$

where A is a constant. The activity of copper (I) ions is determined by the fast pre-equilibrium of reaction (a) when

$$B a_{Cu2+} \exp\left(-\frac{\beta F}{RT}\eta\right) = B' a_{Cu+} \exp\left[\frac{(1-\beta)F}{RT}\eta\right] \qquad (122)$$

$$a_{Cu+} = \frac{B}{B'} \exp\left(-\frac{F}{RT}\eta\right) \qquad (123)$$

where B and B' are constants. Combining Eq. (121) and Eq. (123) gives for the exponential part of the current $\exp[-(1+\beta)F\eta/RT]$. Thus $\alpha_c = 1 + \beta \approx 1.5$. A similar treatment may be made for the anodic reactions.

In general, α_c and α_a are given by

$$\alpha_c = \frac{\gamma}{\nu} + \beta$$

$$\alpha_a = \frac{n-\gamma}{\nu} - \beta$$

where the overall reaction has n electrons, γ charge transfer steps before the rate-determining step, which is repeated ν times (ν is the stoichiometric number). It is to be noted that the sum of the cathodic and anodic transfer coefficients is n/ν.

Transference number (n_i). The number of moles of a given ionic species transferred in the direction of the positive current for the passage of 1 faraday of electricity.

For a 1:1 electrolyte that dissociates normally, the transference number and the **transport number** are identical. When more than two ions are present in solution, the situation is different. In a solution of sulphuric acid which contains the species H^+, HSO_4^- and SO_4^{2-}, the total quantity of electricity passed through the solution will be made up of three contributions in the proportions $\Lambda_{H^+} c_{H^+}$ for the cation, and $2\Lambda_{SO4,2-} c_{SO4,2-}$ and $\Lambda_{HSO4,-} c_{HSO4,-}$ for the anions. The transport number definition is not applicable, but changes around the electrodes enable the transference numbers of the separate ions to be determined. An amount $\Lambda_{H^+} c_{H^+}$ mole of hydrogen ions is transferred towards the cathode, while $\Lambda_{HSO4-} c_{HSO4-}$ mole of hydrogen ions move towards the anode and so

$$n_{H^+} = \frac{\Lambda_{H^+} c_{H^+} - \Lambda_{HSO4-} c_{HSO4-}}{\Lambda_{H^+} c_{H^+} + \Lambda_{HSO4-} c_{HSO4-} + 2\Lambda_{SO4,2-} c_{SO4,2-}}$$

Transient methods. Measurement of an electrochemical variable (current or potential) following a perturbation. The different methods may be classed in terms of the parameter which is controlled: potential (voltammetry, chronoamperometry, chronogalvammetry, chronocoulometry) or current (chronopotentiometry). The perturbation may be continuously varying or in the form of a step. Some common non-steady-state methods are described in Table 30.

In all transient methods, the effect of the double layer is important and must be taken into account. Indeed, the variation of potential following a pulsed current step is used to measure the capacitance of the double layer (*see* **Capacitance**). One method of filtering out double-layer contributions to transient currents is to use a double pulse adjusted so that the initial, short pulse only charges the double layer and the second pulse brings about the desired electrochemical process. The duration of the initial current pulse is such that $dE/dt = 0$ at the beginning of the second pulse.

The use of galvanic steps introduces the concept of **transition time** as the potential changes between one reaction and another. The coverage of intermediates, oxide layers, etc. is measured conveniently by potential steps from one potential at which the species is stable to one at which it is oxidized or reduced. The flow of current, eventually decaying to zero as the species is consumed, may be integrated to yield the coverage. This method is the basis of the electrochemical measurement of surface area by hydrogen stripping (*see* **Surface area determinations**). Direct measurement of the charge passed (chronocoulometry) as the potential is stepped to a value corresponding to the **diffusion-limited current** may be used to obtain values of the **transfer coefficient** and an electrochemical rate constant. Continuously varying stimuli may be in the form of a ramp (*see* **Linear sweep voltammetry**), or an ac signal (*see* **Ac impedance methods**). **Polarography**, which uses a continuously growing drop, is itself a transient method. In addition, a superimposed waveform

Transient methods

Table 30
Non-steady-state methods for investigating electrochemical reactions

Method	Variable Controlled	Variable Observed	Remarks
Chronogalvammetry (chronoamperometry) (voltammetry)			
Potential step	Potential	Current	Requires small IR drop and correction for capacitance currents
Chain of potential steps. Pulse voltammetry (polarography) (e.g., differential pulse, differential normal pulse, differenced normal pulse, reversed pulse, normal pulse)	Potential	Current	Complex mathematical treatment. Control and analysis aided by computers. Double layer effects removed in differential methods. Kinetic information obtained plus data on intermediates
Linear sweep voltammetry (cyclic voltammetry)	Potential	Current	Useful for first observation of intermediates. Some kinetic information, although mathematical treatment complex
Chronopotentiometry			
Current step	Current	Potential	Basic of the **interruptor technique** for measurement of IR drop
Double current step	Current	Potential	First step is adjusted to charge the double layer
Alternating current methods			
Impedance methods	Potential	ac	Measurement at equilibrium or with dc bias which may vary with time
Faradaic rectification	ac	Phase of ac potential	Powerful method although apparatus complex and interpretation difficult
Chronocoulometry	Potential	Charge	Current response integrated. Intermediates studied

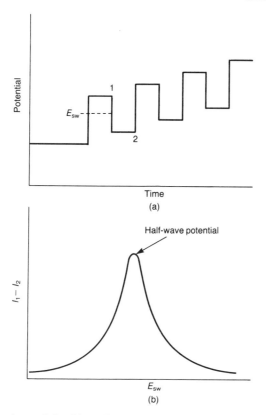

Figure 121. The form of the (a) applied potential and (b) differential current response in square-wave voltammetry

(square wave, sine wave with constant or increasing amplitude) may be applied to the voltage ramp.

The use of microcomputers to provide the waveform and then record and analyse the transients has allowed a greater accuracy in the measurement of half-wave potentials, kinetic parameters, etc., especially in solutions having low concentrations of the electroactive species. For example in square-wave voltammetry, a voltage of the form shown in Figure 121 is applied to the electrode and the current difference $I_1 - I_2$ recorded as a function of the mean potential. The wave is characterized by the amplitude of each pulse, the increment of voltage and the period of the pulse. For a fast scan, these values may be 25 mV, 5 mV and 15 μs, respectively. See Bard and Faulkner (1980).

Transition time. When a constant current is passed through an electrolytic cell, a species which is being discharged will fall in concentration in the region of the electrode. If the species is an adsorbed intermediate, its concentration

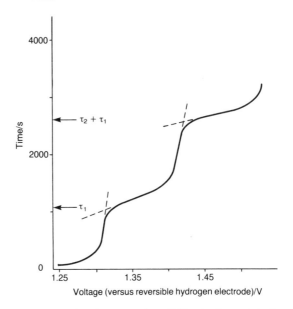

Figure 122. Constant current charging curves for a $NiCo_2O_4$ electrode in 5 mol dm^{-3} KOH at 25°C (current = 1 mA), showing calculation of the transition times for the formation of surface oxides

on the electrode will fall and approach zero. During this process, the potential at the electrode slowly increases as the overpotential increases to account for this fall in concentration. As the surface concentration reaches zero, the potential changes rapidly. The time taken to reach this point is the transition time (τ). For a current density i discharging an ion from solution on a planar electrode, **Fick's laws of diffusion** are followed and lead to the result

$$\tau^{1/2} = \frac{nF}{2i} c_0(\pi D)^{1/2}$$

where D is the diffusion coefficient. The square root of the transition time is thus proportional to the initial concentration (c_0). The current must be adjusted to give a transition time of the order of seconds, to avoid convective mixing, and the rise in potential is monitored on an oscilloscope. Very low concentrations may be measured in this way.

The use of transition times to investigate intermediates is seen in the example of oxygen evolution on $NiCo_2O_4$ in alkaline solution. A constant current applied to such an electrode at a rest potential of 1.00 V leads to the potential–time profile of Figure 122. The transitions at 1.31 V and 1.40 V show the successive oxidations of surface ions ($Ni^{2+} \rightarrow Ni^{3+}$ and $Co^{2+} \rightarrow Co^{3+}$). A final transition, which is concomitant with oxygen evolution, is thought to be $M^{3+} \rightarrow M^{4+}$, where M is nickel or cobalt.

Transport number (t_i). The fraction of the current carried by a given ion. The ions of an electrolyte do not all migrate at the same rate; unequal radii, differences of solvation and different ionic charges give rise to different **mobility** values and hence transport numbers.

$$t_i = \frac{j_i}{\sum_i j_i} = \frac{F|z_i|c_i v_i}{F \sum_i |z_i|c_i v_i} = \frac{v_i}{\sum_i v_i} \quad (124)$$

$$(c_+ z_+ = c_- z_-)$$

that is

$$t_+ = \frac{v_+}{v_+ + v_-} = \frac{u_+}{u_+ + u_-} = \frac{v_+ \Lambda_+}{v_+ \Lambda_+ + v_- \Lambda_-} \quad (125)$$

and

$$t_+ + t_- = 1$$

As the concentration approaches zero, the limiting value of the transport number is approached

$$t_\pm^o = \frac{v_\pm \Lambda_\pm^o}{v_\pm \Lambda_\pm^o + v_\mp \Lambda_\mp^o} \quad (126)$$

A knowledge of the transport numbers enables the calculation of the separate contributions of anions and cations to the total **molar conductivity**. In mixed electrolytes, the relative concentrations influence the quantity of electricity carried by the separate species and the concentration terms in Eq. (124) do not cancel. Under these circumstances, it is customary to calculate the **transference number**.

Since the transport number is related to the molar conductivity, which itself varies with concentration according to a square root relationship (*see* **Conductance equations**), it follows that the transport number should vary with concentration according to an equation of the form

$$t_\pm = t_\pm^o + Bc^{1/2} \quad (127)$$

where B is a constant which for most ions is very small.

The Fuoss–Onsager equation for the transport number of a 1:1 electrolyte is

$$t_\pm = t_\pm^o + \frac{(t_\pm^o - 0.5)\beta c^{1/2}}{\Lambda^o(1 + \kappa a)} \quad (128)$$

where β is the electrophoretic parameter. It follows that at very low concentrations the transport number varies linearly with the square root of the concentration

$$\left(\frac{\partial t_\pm}{\partial c^{1/2}}\right)_{c \to 0} = \frac{(t_\pm^o - 0.5)\beta}{\Lambda^o} \quad (129)$$

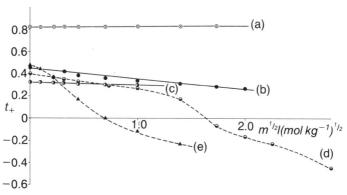

Figure 123. Variation of the transport number of selected cations with concentration: (a) H^+/HCl; (b) $Zn^{2+}/Zn(ClO_4)_2$; (c) $Li^+/LiCl$; (d) $Zn^{2+}/ZnBr_2$; (e) Cd^{2+}/CdI_2

Thus transport numbers of greater than 0.5 should increase with increasing concentration and those less than 0.5 should decrease, whereas values of approximately 0.5 should be independent of concentration.

In water, the experimental results are not always adequately described by these relationships; in other non-aqueous solvents, they fail completely. For very simple completely dissociated electrolytes, the square root relationship is valid (see Figure 123), including the transport number of the potassium ion in potassium chloride dissolved in water, methanol, methanol/water, ethylene glycol and formamide. Some electrolytes (e.g., the halides of cadmium and zinc), however, show very marked deviations, the transport number of the cation becoming negative at high concentrations (see Figure 123). Such results can be attributed to the formation of complex ions: for example

$$Cd^{2+} + Cl^- \rightleftharpoons CdCl^+ \qquad CdCl^+ + 3Cl^- \rightleftharpoons CdCl_4^{2-}$$

Since the complex ion is an anion, it will travel towards the positive electrode thereby transporting cadmium to the positive rather than to the negative electrode. As the sum of the transport numbers must be unity, it follows that that of the halide ion must exceed unity (see also **Colloidal electrolytes**).

The low value for the transport number of the lithium ion compared with values for the other alkali metal ions is due to the more extensive solvation of the very small lithium ion. The large value for the solvated proton is due to the ease of proton transfer between the ion and the solvent (see also **Molar ionic conductivity**). From transport number determinations in non-aqueous solvents, the molar ionic conductivities of ions can be calculated and compared (see Appendix, Table 12).

Determination of transport numbers.
1) Hittorf method. During electrolysis, the concentration of electrolyte in the

vicinity of the anode and cathode changes as a result of reactions at the electrodes and the migration of ions through the solvent. Transport numbers of the ions can be calculated from the measured concentration changes. Consider the electrolysis of an electrolyte, using inert electrodes, in a cell which is arbitrarily divided into three compartments (anode, intermediate and cathode). The following changes occur for the passage of Q coulombs of electricity

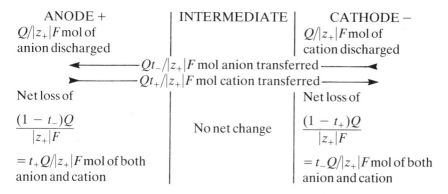

ANODE +	INTERMEDIATE	CATHODE −
$Q/|z_+|F$ mol of anion discharged		$Q/|z_+|F$ mol of cation discharged
←——— $Qt_-/|z_+|F$ mol anion transferred ———		
——— $Qt_+/|z_+|F$ mol cation transferred ———→		
Net loss of		Net loss of
$\dfrac{(1-t_-)Q}{|z_+|F}$	No net change	$\dfrac{(1-t_+)Q}{|z_+|F}$
$= t_+ Q/|z_+|F$ mol of both anion and cation		$= t_- Q/|z_+|F$ mol of both anion and cation

From a knowledge of the concentration changes around both electrodes and the total quantity of electricity passed, using a coulometer in series with the Hittorf cell (see Figure 124), the transport numbers of the ions can be calculated.

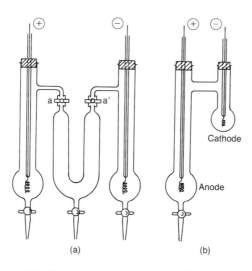

Figure 124. Hittorf cells for transport number determination

Transport number

It is preferable to avoid systems that involve the evolution of a gas at the electrode as this will result in stirring and intermixing of the various parts of the solution. This can be achieved by using electrodes which consist of the metal that is the cation of the electrolyte under study, the metal then dissolves from the anode and is deposited on the cathode during electrolysis: for example, silver electrodes in a silver nitrate solution. Thus for the passage of Q coulombs of electricity

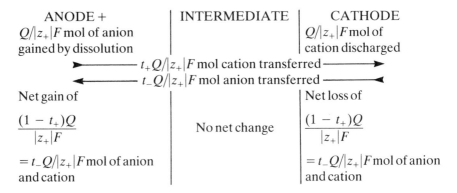

In this situation, there is an increase in the concentration of electrolyte in the vicinity of one electrode and a decrease at the other, whence

$$t_- = \frac{\text{loss of electrolyte at cathode}}{\text{total quantity of electricity passed}}$$

The values of the transport numbers obtained by this method are not in good agreement with those determined by independent methods because most ions are solvated and the solvent is transported with the migrating ions. This problem may be overcome by the addition of a reference substance (e.g., sucrose or raffinose) which is assumed to remain unchanged during electrolysis. The change in concentration of the electrolyte is now calculated with reference to a unit mass of reference substance. The application of the Hittorf method, which has also been used in non-aqueous solvents, is limited by two main factors: (a) at least one and, preferably, both electrodes must be reversible; (b) extreme accuracy is needed in the analysis of the solution before and after electrolysis.

2) Moving boundary method. This method depends on the observation of the movement, under a constant applied current, of the boundary between the solution under study—the 'leading' solution—and a 'following' or 'indicator' solution. The conditions necessary for a sharp boundary are (a) the electrolytes must have a common anion when the transport number of the cation is

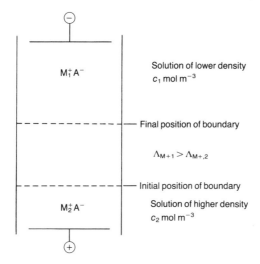

Figure 125. Principle of the moving boundary method

being observed and *vice versa*; (b) the conductivity of the indicator ion must be lower than that of the leading ion; (c) the solution of lower density must be above the solution of higher density; and (d) the concentrations of the two solutions must be in the ratio of their transport numbers (*see* Figure 125).

For Q coulombs of electricity passed, $t_i Q$, is the quantity of electricity carried by ions i and $t_i Q/|z_i|F$ mol migrate. If the concentration of the moving electrolyte is c mol m^{-3}, the volume occupied by this amount of electrolyte is $t_i Q/|z_i|Fc$ m^3; this is the volume V swept out by the boundary when Q coulombs of electricity are passed. Thus

$$V = \frac{t_i Q}{|z_i|Fc}$$

or

$$t_i = \frac{|z_i|FcV}{Q} = \frac{|z_i|FcV}{It} \qquad (130)$$

Thus from a knowledge of the volume swept out by the boundary in time t seconds when the current flowing is I amps, the transport number of the ions can be calculated.

Since the boundary moves at a constant velocity, Eq. (130) applies to both the leading and the following ions thus

$$t_{M+,1} = \frac{Fc_1 V}{It} \quad \text{and} \quad t_{M+,2} = \frac{Fc_2 V}{It}$$

Transport number

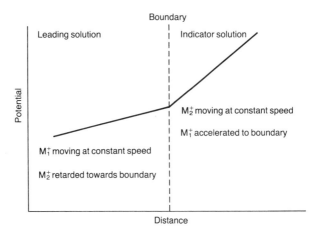

Figure 126. Conditions for the maintenance of a sharp boundary

and hence

$$\frac{t_{M+,1}}{t_{M+,2}} = \frac{c_1}{c_2} \tag{131}$$

that is c_i/t_i = constant. Although it appears from Eq. (131) (Kohlrausch's regulating function) that a previous knowledge of the ratio of the transport numbers of the leading and following ions is required before the concentration of the following ion can be calculated, in practice it is only necessary for Eq. (131) to hold within 10 percent.

Since $\Lambda_{M+,2} < \Lambda_{M+,1}$, the potential gradients in the two solutions must differ if the ions M_1^+ and M_2^+ are to move with the same speed and maintain a sharp boundary. For a given current, the potential gradient in the indicator solution is steeper than it is in the leading solution (*see* Figure 126).

Any M_1^+ ions lagging behind the boundary (i.e. in the indicator solution) will be accelerated towards the boundary by the higher potential gradient. Similarly any M_2^+ ions in front of the boundary will be slowed down. Very accurate results have been obtained by MacInnes and Longsworth with the apparatus illustrated in Figure 127.

3) E.m.f. method. The transport number of an ion can be obtained by comparing the e.m.f. of **concentration cells** with transport (Eq. (19)) and without transport (Eq. (20)), thus

$$t_i = \frac{E_{\text{transport}}}{E}$$

This method assumes that the transport numbers are independent of concentration. *See also* **Isotachophoresis**.

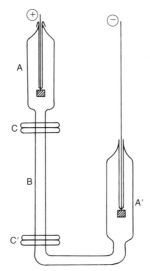

Figure 127. Moving boundary apparatus. Electrode vessels A,A' end in glass discs C,C' which can be rotated into or out of line with the graduated tube B. The discs are used to form a sharp boundary between the leading and indicator solutions

Further reading
R. L. Kay, Transference number measurements, in *Techniques of Electrochemistry*, vol. 2, E. Yeager and A. J. Salkind (ed.), Wiley–Interscience, New York, p. 61 (1973)

M. Spiro, Transference numbers, in *Physical Methods of Chemistry*, vol. IIA, A. Weissberger and B. W. Rossiter (ed.), Wiley–Interscience, New York, p. 205 (1971)

J. Barthel, R. Wachter and H. J. Gores, Temperature dependence of conductance of electrolytes in non-aqueous solutions, in *Comprehensive Treatise of Electrochemistry*, vol. 13, J. O'M. Bockris *et al.* (ed.), Plenum Press, New York (1984)

V

Volmer reaction. *See* **Hydrogen electrode reactions.**

Volt (V; dimensions: $\varepsilon^{-1/2} m^{1/2} l^{1/2} t^{-1}$; units: $V = kg\, m^2\, s^{-3}\, A^{-1} = J\, A^{-1}\, s^{-1}$). The SI unit of electromotive force defined as the difference of potential required to make a current of 1 amp flow through a resistance of 1 ohm.

Voltaic cell. *See* **Reversible galvanic cell.**

Voltammetric titration. *See* **Electrometric titrations.**

Voltammetry. *See* **Transient techniques.**

Volta potential. *See* **Outer potential.**

W

Walden's Law. *See* **Molar ionic conductivity.**

Weston cell. *See* **Standard cell.**

Wien effects. *See* **Conductance at high field strengths.**

Working electrode. *See* **Three-electrode system.**

Z

Zeta-potential (ζ). The potential difference, measured in the liquid, between the shear plane (i.e. the outer limit of that part of the double layer which is fixed to the phase moving relative to the liquid) and the bulk of the liquid beyond the limits of the **electrical double layer**. The ζ-potential is quite different from the contact and Nernst potentials between the two phases (*see* Figures 128 and 129). The ζ-potential cannot be measured directly, but can be calculated from measurements of the **electrophoretic mobility** of particles using the Smoluchowski equation

$$u = \frac{v}{X} = \frac{\varepsilon \zeta}{4\pi\eta} = \frac{\varepsilon_0 \varepsilon_r \zeta}{\eta} \tag{132}$$

In water at 298 K

$$\zeta/\text{mV} = 12.85 \times (u/10^{-8} \, \text{m}^2 \, \text{s}^{-1} \, \text{V}^{-1})$$

Eq. (132) is applicable to a non-conducting particle of 'easy' shape provided that the radius of curvature is much greater than the thickness of the electrical double layer (for particles of radius a, $\kappa a > 100$ where κ is the reciprocal thickness of the double layer). The derivation of Eq. (132) involves the assumptions that the conductivity, permittivity and viscosity have the same values within the double layer as in the bulk liquid, that the applied electric field can be added to the field of the double layer and that the velocity gradient begins at the particle surface. Other equations have been advanced for special situations.

The surface charge density in the diffuse layer for a surface in an electrolyte of single valence can be calculated from the ζ-potential by

$$\sigma_\text{D} = \left(\frac{N_\text{A} \varepsilon k T}{500\pi}\right)^{1/2} c^{1/2} \sinh \frac{ze\zeta}{kT} \tag{133}$$

$$\sigma_\text{D}/\text{Cm}^{-2} = 3.713 \times 10^{-3} (c/\text{mol m}^{-3})^{1/2} \sinh \left[(\zeta/\text{mV})/51.3\right]$$

The variation of the charge carried by a particle with ionic strength depends on the valence type of the electrolyte and on the presence of ions that may undergo specific interaction or adsorption with the surface. In general, with increasing concentration of electrolyte, there is an increase in the negative ζ-potential (of a negatively charged surface), which passes through a

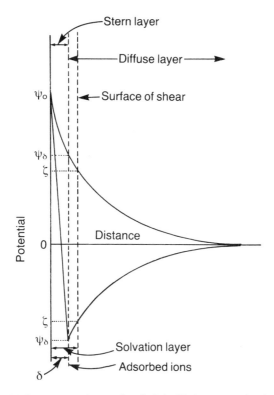

Figure 128. Potential decay curves for an electrical double layer associated with a particle of surface potential ψ_0. The lower curve results from strong adsorption within the Stern layer

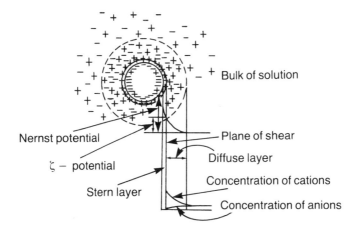

Figure 129. Schematic representation of charge distribution around a particle with a net negative charge

296 Zeta-potential

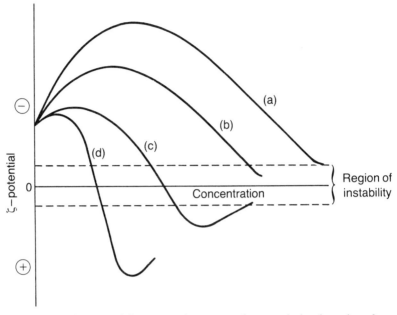

Figure 130. Typical ζ-potential–concentration curves for negatively charged surfaces in electrolytes of different valence type: (a) 1:1 (potassium chloride); (b) 2:1 (barium chloride); (c) 3:1 (cerium chloride); (d) 4:1 (thorium nitrate)

maximum value and thereafter decreases (see Figure 130). The initial increase of charge is often referred to as the 'charging process'. The decrease of the ζ-potential is determined by the nature and concentration of the ion of opposite sign to that of the particle (i.e. cation for a negatively charged surface). In contrast, the charge and nature of the anion has an insignificant effect on the shape or position of the curves. Thus the concentration of electrolyte required for the particle or surface to attain a zero charge (i.e. **isopotential point**) is greatest for a univalent ion and decreases with increasing valence. Thus the sign of the charge carried by a negatively charged particle may be completely reversed in the presence of a very low concentration of a thorium salt solution; this is the charge-reversal concentration.

The ζ-potential is responsible for the stability of hydrophobic sols; at very low values, the forces of attraction are greater than the electrical forces of repulsion and the sol becomes unstable and is irreversibly flocculated. In contrast, the ζ-potential plays a very minimal role in determining the stability of hydrophilic sols.

The ζ-potential of a surface also depends on the pH of the surrounding electrolyte and on the presence of surfactants in the electrolyte. See also **Surface charge**.

Further reading
M. Bier (ed.), *Electrophoresis, Theory, Methods and Applications*, Academic Press, New York (1959)

Zinc–air cell. *See* **Metal–air batteries.**

Zinc electrometallurgy. Zinc is prepared electroanalytically from an acid solution of purified zinc sulphate, using aluminium sheets as cathodes and anodes of pure lead. The anodic reaction is the liberation of oxygen, and so the concentration of free sulphuric acid tends to increase continuously. To counter this the electrolyte is circulated, and the more acid solutions are returned to the leaching process.

The standard electrode potential of zinc is -0.76 V, so it might be thought that hydrogen would be evolved preferentially at the cathode. This must be prevented, not only to maintain a high current efficiency, but also because the loss of hydrogen ions from the solution in contact with the cathode would result in a local alkalinity which would lead to a spongy and unsatisfactory deposit. Fortunately it is prevented by the high hydrogen overpotential at a zinc surface. A low working temperature of about 35°C and a high current density (600 A m^{-2}) favour this high overpotential.

The electrolyte must be specially purified because (1) the lead anodes are attacked by an electrolyte containing more than about 50 mg dm^{-3} of chloride ions and (2) almost any metal present in the electrolyte would be deposited on the cathode, and would then tend to lower the necessary high hydrogen overpotential. Germanium is especially deleterious, even in amounts of 0.1 ppm. The current efficiency is about 90 percent, an inevitable loss being the chemical attack of the zinc by the acid electrolyte.

Zinc–carbon battery. *See* **Leclanché cell.**

REFERENCES AND TEXTBOOKS

M. S. Antelman and F. J. Harris, *The Encyclopedia of Electrode Potentials*, Plenum Press, New York (1982)

A. J. Bard, *Encyclopedia of Electrochemistry of the Elements*, vols. 1–14, Marcel Dekker (1973–81)

A. J. Bard and L. R. Faulkner, *Electrochemical Methods, Fundamentals and Applications*, John Wiley, New York (1980)

J. O'M. Bockris (ed.), *Modern Aspects of Electrochemistry*, Butterworths, London (1980–)

J. O'M. Bockris and A. K. N. Reddy, *Modern Electrochemistry*, vols. 1 and 2, Plenum Press, New York (1973)

J. O'M. Bockris, B. E. Conway and E. Yeager (ed.), *Comprehensive Treatise of Electrochemistry*, Plenum Press, New York:
 vol. 1, *The Double Layer* (1980)
 vol. 2, *Electrochemical Processing*, with R. E. White (1981)
 vol. 3, *Electrochemical Energy Conversion and Storage*, with R. E. White (1981)
 vol. 4, *Electrochemical Materials Science*, with R. E. White (1981)
 vol. 5, *Thermodynamic and Transport Properties of Aqueous and Molten Electrolytes* (1982)
 vol. 6, *Electrodics: Transport*, with S. Sarangapani (1982)
 vol. 7, *Kinetics and Mechanisms of Electrode Processes*, with R. E. White and S. V. M. Khan (1983)
 vol. 8, *Experimental Methods in Electrochemistry*, with R. E. White (1984)
 vol. 9, *Electrodics*, with S. Sarangapani (1984)
 vol. 10, *Bioelectrochemistry*, with S. Srinivasan and Y. A. Chizmadzhev (1984)

B. E. Conway, Ionic hydration in chemistry and biophysics, in *Studies in Physical and Theoretical Chemistry*, vol. 12, Elsevier, Amsterdam (1981)

A. K. Covington and T. Dickinson (ed.), *Physical Chemistry of Organic Solvent Systems*, Plenum Press, London (1973)

T. R. Crompton, *Small Batteries*, Macmillan, London (1982)

C. W. Davies, *Ion Association*, Butterworths, London (1962)

P. Delahay and C. W. Tobias (ed.), *Advances in Electrochemistry and Electrochemical Engineering*, John Wiley, New York (1960–)

M. Dole, *Principles of Experimental and Theoretical Electrochemistry*, McGraw-Hill, New York (1935)

References and textbooks

U. R. Evans, *The Corrosion and Oxidation of Metals*, Edward Arnold, London (1960)

I. Fried, *The Chemistry of Electrode Processes*, Academic Press, London (1973)

S. Glasstone, *Introduction to Electrochemistry*, van Nostrand, New York (1942)

H. S. Harned and B. B. Owen, *The Physical Chemistry of Electrolytic Solutions*, Reinhold, New York (1958)

D. Inman and D. G. Lovering (ed.), *Ionic Liquids*, Plenum Press, New York (1981)

D. J. G. Ives and G. J. Janz (ed.), *Reference Electrodes: Theory and Practice*, Academic Press, London (1961)

A. M. James, *A Dictionary of Thermodynamics*, Macmillan, London (1976)

A. M. James and F. E. Prichard, *A Text-book of Practical Physical Chemistry*, Longman, London (1977)

G. J. Janz and R. P. T. Tomkins (ed.), *Non-aqueous Electrolytes Handbook*, vol. 1, Academic Press, New York (1972)

G. J. Janz and R. P. T. Tomkins (ed.), *Non-aqueous Electrolytes Handbook*, vol. 2, Academic Press, New York (1973)

Kirk–Othmer Encyclopedia of Chemical Technology, 3rd edn. Wiley–Interscience, New York (1981)

J. Koryta (ed.), *Medical and Biological Applications of Electrochemical Devices*, Wiley–Interscience, London (1980)

U. Landau, E. Yeager and D. Kortan, *Electrochemistry in Industry. New Directions*, Plenum Press, New York (1982)

D. G. Lovering (ed.), *Molten Salt Technology*, Plenum Press, New York (1982)

D. G. Lovering and R. J. Gale, *Molten Salt Techniques*, vol. 1, Plenum Press, New York (1983)

G. Milazzo and M. Blank (ed.), *Biochemistry, I. Biological Redox Reactions*, Plenum Press, New York (1983)

R. Parsons, in *Advances in Electrochemistry and Electrochemical Engineering*, vol. 1, Wiley–Interscience, New York (1961)

D. Pletcher, *Industrial Electrochemistry*, Chapman and Hall, London (1982)

R. A. Robinson and R. H. Stokes, *Electrolyte Solutions*, Butterworths, London (1970)

E. Yeager and A. J. Salkind, *Techniques of Electrochemistry*, vol. 1, Wiley–Interscience, New York (1972)

E. Yeager and A. J. Salkind, *Techniques of Electrochemistry*, vol. 2, Wiley–Interscience, New York (1973)

E. Yeager and A. J. Salkind, *Techniques of Electrochemistry*, vol. 3, Wiley–Interscience (1978)

ELECTROCHEMICAL JOURNALS

General electrochemistry

Journal of the Electrochemical Society	Vol. 1–	1854
Journal of Electroanalytical Chemistry and Interfacial Electrochemistry	Vol. 1–	1960
Electrochimica Acta	Vol. 1–	1956
Zeitschrift für Elektrochemie (GDR)	Vol. 1–	1954
Elektrokhymia (USSR)	Vol. 1–	1965
Journal of Applied Electrochemistry	Vol. 1–	1971
Denki Kagaku (Japan)	Vol. 1–	1933
Journal of Electrochemical Society, India	Vol. 1–	1951
Transactions of the Society for Advancement of Electrochemical Science and Technology (India)	Vol. 1–	1966

Specialist topics

Plating (journal of the American Electroplaters' Society)	Vol. 1–	1914
which became Plating and Surface Finishing	Vol. 1–	1975
Electrophoresis	Vol. 1–	1980
Journal of Power Sources	Vol. 1–	1976
Journal of the Polarographic Society	Vol. 1–	1960
Review of Polarography (journal of the Polarographic Society, Japan)	Vol. 1–	1960
Journal of Applied Microbiology and Biotechnology (translation from Russian) Faraday Press	Vol. 1	1965
Enzyme and Microbial Technology (IPC Science and Technology)	Vol. 1	1979
Enzyme Engineering (Plenum Press, London)	Vol. 1	1972
Biotechnology and Bioengineering (previously Journal of Biochemical and Microbial Technology and Engineering)	Vol. 4 Vol. 1	1962 1959
Applied Biochemistry and Biotechnology (Humana Press Inc., NJ)	Vol. 6	1980
(previously Journal of Solid-Phase Biochemistry)	Vol. 1	1976
Bioelectrochemistry and Bioenergetics (Basle)	Vol. 1	1974

Reviews

Advances in Electroanalytical chemistry A. J. Bard (ed.) (Marcel Dekker, New York)	Vol. 1–	1966
Ion Selective Electrode Reviews (Pergamon, Oxford)	Vol. 1–	1979
Electrochemistry (Special Publications Royal Society of Chemistry, London)	Vol. 1–	1971

Appendix

TABLES OF USEFUL DATA

Table 1
Recommended values of physical constants

Physical constant	Symbol	Value
Acceleration due to gravity	g	$9.81 \, \text{m s}^{-2}$
Avogadro constant	N_A	$6.02252 \times 10^{23} \, \text{mol}^{-1}$
Bohr magneton	μ_B	$9.2732 \times 10^{-24} \, \text{A m}^2 (\text{J T}^{-1})$
Boltzmann constant	k	$1.38054 \times 10^{-23} \, \text{J K}^{-1}$
Charge-to-mass ratio	e/m	$1.758796 \times 10^{11} \, \text{C kg}^{-1}$
Curie	Ci	37.0×10^9 disintegrations per second
Electronic charge	e	$1.60210 \times 10^{-19} \, \text{C}$
Faraday constant	F	$9.64870 \times 10^4 \, \text{C mol}^{-1}$
Gas constant	R	$8.3143 \, \text{J K}^{-1} \text{mol}^{-1}$
Gravitational constant	G	$66.7 \times 10^{-12} \, \text{m}^3 \text{kg}^{-1} \text{s}^{-2}$
'Ice-point' temperature	T_{ice}	$273.150 \, \text{K}$
Molar volume of ideal gas at STP	V_m	$2.24136 \times 10^{-2} \, \text{m}^3 \text{mol}^{-1}$
Permeability of a vacuum	μ_0	$4\pi \times 10^{-7} \, \text{kg m s}^{-2} \text{A}^{-2} (\text{H m}^{-1})$
Permittivity of a vacuum	ε_0	$8.854185 \times 10^{-12} \, \text{kg}^{-1} \text{m}^{-3} \text{s}^4 \text{A}^2 (\text{F m}^{-1})$
Planck constant	h	$6.6256 \times 10^{-34} \, \text{J s}$
Rydberg constant	R_∞	$1.0973731 \times 10^7 \, \text{m}^{-1}$
Standard pressure, atmosphere	P	$101325 \, \text{N m}^{-2}$
Stefan–Boltzmann constant	σ	$5.6697 \times 10^{-8} \, \text{W m}^{-2} \text{K}^{-4}$
Triple point of water		$273.16 \, \text{K}$ (exactly)
Unified atomic mass constant	m_u	$1.66043 \times 10^{-27} \, \text{kg}$
Velocity of light in a vacuum	c	$2.997925 \times 10^8 \, \text{m s}^{-1}$
Wien's radiation law	$\lambda_{max} \times T$	$2.8978 \times 10^{-3} \, \text{m K}$

Tables of useful data

Table 2
Relative permittivity (dielectric constants) of solvents (at 20°C unless otherwise stated)

Solvent	ε_r
N-methyl acetamide (40°C)	165
Hydrocyanic acid	115
Formamide	109
Water	80.36
Formic acid	57
Ethylene glycol	41.2
Ethanolamine (25°C)	37.7
Acetonitrile (25°C)	37.
Methanol	33.7
Benzonitrile (25°C)	25.2
Ethanol (25°C)	24.5
Ammonia (-33°C)	22.0
Acetone	21.45
Cyclohexanone (25°C)	18.3
Sulphur dioxide (0°C)	15.4
Ethylenediamine (25°C)	14.2
Pyridine (25°C)	12.3
Ethylene chloride (25°C)	10.2
Methylamine	10.0
Acetic acid	6.1
Chloroform	4.8
Diethyl ether	4.38
Benzene	2.29
Dioxan	2.24

Table 3
Standard (reduction) potentials and the temperature coefficients at 298 K in aqueous solution

Electrode	$E^{\ominus}$/V	$(\partial E^{\ominus}/\partial T)$/mV K^{-1}
$Li^+\|Li$	−3.045	−0.59
$K^+\|K$	−2.925	−1.07
$Ba^{2+}\|Ba$	−2.90	−0.40
$Sr^{2+}\|Sr$	−2.89	−0.23
$Ca^{2+}\|Ca$	−2.87	−0.21
$Na^+\|Na$	−2.7141	−0.75
$Mg^{2+}\|Mg$	−2.375	+0.18
$Al^{3+}\|Al$	−1.66	+0.53
$OH^-\|H_2,Pt$	−0.8281	−0.80
$Zn^{2+}\|Zn$	−0.7631	+0.10
$Fe^{2+}\|Fe$	−0.441	+0.05
$Cd^{2+}\|Cd$	−0.4019	−0.09
$SO_4^{2-}\|PbSO_4,Pb$	−0.3553	−0.99
$I^-\|AgI,Ag$	−0.1524	−0.33
$Pb^{2+}\|Pb$	−0.1288	−0.38
$H_3O^+\|H_2,Pt$	0	0
$Br^-\|AgBr,Ag$	+0.0711	−0.49
$Cl^-\|AgCl,Ag$	+0.2224	−0.66
$Cl^-\|Hg_2Cl_2,Hg$	+0.2681	−0.31
$Cu^{2+}\|Cu$	+0.337	+0.01
$I^-\|I_2,Pt$	+0.5346	−0.13
$Fe^{3+},Fe^{2+}\|Pt$	+0.771	+1.19
$Hg_2^{2+}\|Hg$	+0.796	−0.31
$Ag^+\|Ag$	+0.7991	−1.00
$Br^-\|Br_2,Pt$	+1.065	−0.61
$H^+\|O_2,Pt$	+1.229	−0.85
$Cl^-\|Cl_2,Pt$	+1.3595	−1.25

Tables of useful data

Table 4
Standard (reduction) potentials at 298 K (except liquid ammonia 265 K) $E^{\ominus}_{H^+,H_2O} = 0$ V

Electrode	Water	Methanol	Acetonitrile	Formamide	Hydrazine	Ammonia
$Li^+\|Li$	−3.04	−3.13	−3.09		−3.11	−3.23
$Ca^{2+}\|Ca$	−2.87		−2.61		−2.82	−2.73
$Na^+\|Na$	−2.71	−2.76	−2.73		−2.74	−2.84
$Zn^{2+}\|Zn$	−0.76	−0.77	−0.60	−0.83	−1.32	−1.52
$Cd^{2+}\|Cd$	−0.40	−0.46	−0.33	−0.48	−1.01	−1.19
$I^-\|AgI,Ag$	−0.152	−0.318				
$Pb^{2+}\|Pb$	−0.13	−0.23	+0.02	−0.26	−0.56	−0.67
$H^+\|H_2,Pt$	0	−0.03	+0.14	−0.07	−0.91	−0.99
$Br^-\|AgBr,Ag$	+0.071	−0.14				
$Cl^-\|AgCl,Ag$	+0.2224	−0.010		+0.204		
$Cu^{2+}\|Cu$	+0.34	+0.31	−0.24	+0.21		−0.56
$Ag^+\|Ag$	+0.799	+0.73	+0.37		−0.14	−0.16
$Cl^-\|Cl_2,Pt$	+1.36	+1.09	+0.72			+1.04

Table 5
Standard redox potentials at 298 K

Electrode system	$E^{\ominus}/V$
$Cr^{3+}\|Cr^{2+}$	−0.41
$Tl^{3+}\|Tl^{2+}$	−0.37
$Sn^{4+}\|Sn^{2+}$	+0.15
$Cu^{2+}\|Cu^+$	+0.159
$Fe(CN)_6^{3-}\|Fe(CN)_6^{4-}$	+0.356
Quinone|hydroquinone, $a_{H^+} = 1$	+0.6995
$Fe^{3+}\|Fe^{2+}$	+0.771
$Hg^{2+}\|Hg_2^{2+}$	+0.91
$Cr_2O_7^{2-}\|Cr^{3+}$	+1.36
$MnO_4^-, H^+\|Mn^{2+}, a_{H^+} = 1$	+1.52
$Ce^{4+}\|Ce^{3+}$	+1.61
$Co^{3+}\|Co^{2+}$	+1.81

Tables of useful data

Table 6
Standard redox potentials for some biological half reactions at 298 K and at pH = 7.0

System	Half cell reaction	$E^{\ominus\prime}/V$
Acetate/pyruvate	$CH_3COOH + CO_2 + 2H^+ + 2e \rightarrow CH_3COCOOH + H_2O$	−0.70
Fe^{3+}/Fe^{2+} (ferredoxin)	$Fe^{3+} + e \rightarrow Fe^{2+}$	−0.432
H^+/H_2	$2H^+ + 2e \rightarrow H_2(g)$	−0.421
$NADP^+/NADPH$	$NADP^+ + 2H^+ + 2e \rightarrow NADPH + H^+$	−0.324
$NAD^+/NADH$	$NAD^+ + 2H^+ + 2e \rightarrow NADH + H^+$	−0.320
$FAD/FADH_2$	$FAD + 2H^+ + 2e \rightarrow FADH_2$	−0.219
Acetaldehyde/ethanol	$CH_3CHO + 2H^+ + 2e \rightarrow C_2H_5OH$	−0.197
Pyruvate/lactate	$CH_3COCOOH + 2H^+ + 2e \rightarrow CH_3CHOHCOOH$	−0.185
Oxaloacetate/malate	$\begin{array}{l} CH_2COOH \\ \mid \\ COCOOH \end{array} + 2H^+ + 2e \rightarrow \begin{array}{l} CH_2COOH \\ \mid \\ CHOHCOOH \end{array}$	−0.166
MB/MBH_2 [a]	$MB + 2H^+ + 2e \rightarrow MBH_2$	+0.011
Fumarate/succinate	$\begin{array}{l} CHCOOH \\ \parallel \\ CHCOOH \end{array} + 2H^+ + 2e \rightarrow \begin{array}{l} CH_2COOH \\ \mid \\ CH_2COOH \end{array}$	+0.031
Fe^{3+}/Fe^{2+} (myoglobin)	$Fe^{3+} + e \rightarrow Fe^{2+}$	+0.046
Fe^{3+}/Fe^{2+} (cytochrome b)	$Fe^{3+} + e \rightarrow Fe^{2+}$	+0.050
Ubiquinone/ubihydroquinone	$Ub + 2H^+ + 2e \rightarrow UbH_2$	+0.10
Cyt c^{3+}/cyt c^{2+}	$Fe^{3+} + e \rightarrow Fe^{2+}$	+0.254
Cyt a^{3+}/cyt a^{2+}	$Fe^{3+} + e \rightarrow Fe^{2+}$	+0.29
Cyt f^{3+}/cyt f^{2+}	$Fe^{3+} + e \rightarrow Fe^{2+}$	+0.365
Cu^{2+}/Cu^+ (haemocyanin)	$Cu^{2+} + e \rightarrow Cu^+$	+0.540
O_2/H_2O	$O_2(g) + 4H^+ + 4e \rightarrow 4H_2O$	+0.816

[a] MB and MBH_2 represent the oxidized and reduced forms of methylene blue.

Table 7
pH values of operational standard solutions recommended for the calibration of glass electrodes (BS 1647:1983)

	pH values at 25°C	pH values at 37°C
0.1 mol kg^{-1} potassium tetroxalate	1.48	1.49
0.01 mol dm^{-3} hydrochloric acid + 0.09 mol dm^{-3} potassium chloride	2.07	2.08
0.05 mol kg^{-1} potassium hydrogen phthalate	4.005	4.022
0.10 mol dm^{-3} acetic acid + 0.1 mol dm^{-3} sodium acetate [a]	4.644	4.647
0.01 mol dm^{-3} acetic acid + 0.01 mol dm^{-3} sodium acetate [a]	4.713	4.722
0.02 mol kg^{-1} piperazine phosphate	6.26	6.14
0.025 mol kg^{-1} disodium hydrogen phosphate + 0.025 mol kg^{-1} potassium dihydrogen phosphate	6.857	6.828
0.05 mol kg^{-1} TRIS hydrochloride + 0.01667 mol kg^{-1} TRIS [b]	7.648	7.332
0.05 mol kg^{-1} disodium tetraborate (borax)	9.182	9.074
0.025 mol kg^{-1} sodium hydrogen carbonate + 0.025 mol kg^{-1} sodium carbonate	9.995	9.889
Saturated calcium hydroxide	12.43	12.05

[a] Prepared from pure acetic acid, diluted and half-neutralized with sodium hydroxide.
[b] Tris(hydroxymethyl)methane.

Tables of useful data

Table 8
Colour changes and pH range of some acid–base indicators

Indicator	Colour Acid	Colour Alkaline	Approximate pH range	pK_{In}
Methyl violet	Yellow	Violet	0.1–2.0	
Thymol blue	Red	Yellow	1.2–2.8	1.7
Bromophenol blue	Yellow	Blue	2.9–4.6	4.0
Methyl orange	Red	Yellow	3.1–4.4	3.7
Methyl red	Red	Yellow	4.2–6.3	5.1
Bromocresol purple	Yellow	Violet	5.2–6.8	6.3
Bromothymol blue	Yellow	Blue	6.0–7.6	6.3
4-Nitrophenol	Colourless	Yellow	5.6–7.6	7.1
Phenol red	Yellow	Red	6.8–8.4	7.9
Thymol blue	Yellow	Blue	8.0–9.6	8.9
Phenolphthalein	Colourless	Red	8.3–10.0	9.6
Thymolphthalein	Colourless	Blue	8.3–10.5	9.2
Alizarin yellow R	Yellow	Red	10.0–12.0	
Tropaeolin O	Yellow	Orange	11.1–12.7	

Table 9
Redox indicators for volumetric use

Indicator	$E^\ominus/V$ (at pH = 0)
Phenosafranin	0.28
Variamine blue B hydrochloride	0.50
Methylene blue	0.52
Diphenylamine	0.76
Barium diphenylaminesulphonate	0.84
Diphenylaminesulphonic acid (33 percent soln)	0.86
Lissamine green	0.99
1,10-Phenanthroline hydrate	1.08
N-Phenylanthranilic acid	1.08

Tables of useful data

Table 10
Redox indicators for biological use

Indicator	$E^{\ominus\prime}/V$ (at pH = 7.0)
Methyl viologen	−0.440
Benzyl viologen	−0.359
Neutral red	−0.325
Phenosafranin	−0.252
Janus green	−0.225
Methylene blue	+0.011
Lauth's violet	+0.063
Phenolindo-2,6-dichlorophenol	+0.217
Phenol blue	+0.224
Phenolindophenol	+0.227

Table 11
Potentials of some common reference electrodes in aqueous solution at 25°C[a]

Electrode	Half cell reaction	$E°$ (25°C)/V
Hydrogen	$H^+\mid H_2,Pt$	0.0
Saturated **calomel**	$Cl^-\mid Hg_2Cl_2,Hg$	+0.242
Normal **calomel**	$Cl^-\mid Hg_2Cl_2,Hg$ (1 mol dm^{-3})	+0.281
0.1 Normal **calomel**	$Cl^-\mid Hg_2Cl_2,Hg$ (0.1 mol dm^{-3})	+0.334
Silver–silver chloride	$Cl^-\mid AgCl,Ag$	+0.222
Silver–silver bromide	$Br^-\mid AgBr,Ag$	+0.071
Silver–silver iodide	$I^-\mid AgI,Ag$	−0.152
Silver–silver oxide	$OH^-\mid Ag_2O,Ag$	+1.170
Mercury–mercury oxide	$OH^-\mid HgO,Hg$	+0.926
Antimony–antimony oxide	$OH^-\mid Sb_2O_3,Sb$	+0.255
Lead–lead sulphate	$SO_4^{2-}\mid PbSO_4,Pb(Hg)$	+0.965
Quinhydrone	$H^+,QH_2,Q\mid Pt$ (where $Q = C_6H_4O_2$)	+0.699

[a] All electrode potentials are given versus a standard hydrogen electrode except the metal–metal oxide electrodes which are versus a hydrogen electrode in the same electrolyte.

Tables of useful data

Table 12
Limiting molar ionic conductivities ($10^4 \, \Lambda^\circ / \Omega^{-1} \, m^2 \, mol^{-1}$) at 298 K in a range of solvents

	Water	Methanol	Ethanol	Ethylene glycol	Formamide	N-Methyl formamide	Dimethyl formamide	Acetonitrile	Nitromethane	Dimethyl sulphoxide	Sulpholane (303 K)	N-Methyl acetamide
ε_r	80.4	33.7	24.5	41.2	109	182.4	36.7	37	36.7	46.7	43.3	165
H+	349.82	146.2	62.8	27.7	10.8		34.7	9.9		15.0		8.95
Li+	38.68	39.6	17.09	2.11	9.03		25.0	69.3			4.34	6.45
Na+	50.11	45.2	20.32	3.11	10.10	21.6	29.9	76.9		14.2	3.61	8.1
K+	73.52	52.5	23.57	4.62	12.75	22.2	30.8	83.6		15.0	4.04	8.3
Rb+	77.81	55.9	24.76		12.8		32.4	85.6			4.20	
Cs+	77.26	60.9	26.58		13.9	24.3	34.5	87.3		16.6	4.34	
Ag+	61.92	50.1	19.23				35.2	86.2		16.4	4.81	
NH4+	73.55				15.6		38.6				4.98	9.55
NMe4+	44.92	68.7	29.8	2.97	13.38	26.2	39.1	94.5	54.5	19.0	4.28	11.9
NEt4+	32.66	60.5	29.4	2.20	11.02		35.4	84.8	47.7	17.5	3.94	11.5
NPr4+	23.45	46.1	23.05	1.74	8.12		29.0	70.3	39.2	13.8	3.23	9.0
NBu4+	19.47	39.05	19.72	1.51	6.83		25.4	61.4	34.1	11.8	2.76	7.7
NAm4+	17.47	34.85			5.8		22.9	56.0		10.8	2.51	7.25
OH-	198.30											
F-	55.41											
Cl-	76.34	52.35	21.85	3.26	17.12	19.8	55.1	98.7	62.5	24.0	9.29	11.7
Br-	78.14	56.45	23.78	5.07	17.17	21.5	53.6	100.7	62.9	23.6	8.91	12.95
I-	76.84	62.75	26.92	4.98	16.73	22.8	52.3	102.4		23.4	7.22	14.75
NO3-	71.46	61.1	25.65	4.61	17.4		57.3	106.4		26.5		14.65
CH3COO-	40.90	53.1		4.81	11.9							
Pic-	31.4	47.1	25.0	2.21	9.1	13.1	37.4	77.7		16.8	5.32	11.95
SCN-	66.0	62.3	27.4		17.2		59.7	113.4		28.6	9.63	16.1

NMe4+, tetramethylammonium ion; NEt4+, tetraethylammonium ion; NPr4+, tetra-iso-propylammonium ion; NBu4+, tetra-t-butylammonium ion; NAm4+, tetraamylammonium ion; Pic-, picrate ion.